BC Police
Math
Workbook

Math Practice, Multiple Choice

Strategies and Tutorials

COMPLETE
TEST PREPARATION INC.
WWW.TEST-PREPARATION.CA

Copyright © 2016 by Complete Test Preparation Inc. ALL RIGHTS RESERVED.

No part of this book may be reproduced or transferred in any form or by any means, graphic, electronic, or mechanical, including photocopying, recording, web distribution, taping, or by any information storage retrieval system, without the written permission of the author.

Notice: Complete Test Preparation Inc. makes every reasonable effort to obtain from reliable sources accurate, complete, and timely information about the tests covered in this book. Nevertheless, changes can be made in the tests or the administration of the tests at any time and Complete Test Preparation Inc. makes no representation or warranty, either expressed or implied as to the accuracy, timeliness, or completeness of the information contained in this book. Complete Test Preparation Inc. make no representations or warranties of any kind, express or implied, about the completeness, accuracy, reliability, suitability or availability with respect to the information contained in this document for any purpose. Any reliance you place on such information is therefore strictly at your own risk.

The author(s) shall not be liable for any loss incurred as a consequence of the use and application, directly or indirectly, of any information presented in this work. Sold with the understanding, the author is not engaged in rendering professional services or advice. If advice or expert assistance is required, the services of a competent professional should be sought.

The company, product and service names used in this publication are for identification purposes only. All trademarks and registered trademarks are the property of their respective owners. Complete

Test Preparation Inc. is not affiliated with any educational institution.

We strongly recommend that students check with exam providers for up-to-date information regarding test content.

Complete Test Preparation Inc., is not affiliated with any British Columbia Police Organization, or the Justice Institute of BC, who are not involved in the production of, and do not endorse, this product.

Published by
Complete Test Preparation Inc.
Victoria BC Canada

Visit us on the web at
https://www.test-preparation.ca
Printed in the USA

Version 6 Updated August 2025

ISBN-13: 9781772455571

About Complete Test Preparation

Why Us?
The Complete Test Preparation Team has been publishing high quality study materials since 2005, with a catalogue of over 145 titles, in English, French, Spanish and Chinese, as well as ESL curriculum for all levels.

To keep up with the industry changes, we update everything all the time!

And the best part?
With every purchase, you're helping people all over the world improve themselves and their education. So thank you in advance for supporting this mission with us! Together, we are truly making a difference in the lives of those often forgotten by the system.

Charities that we support
https://www.test-preparation.ca/charities-and-non-profits/

You have definitely come to the right place.
If you want to spend your valuable study time where it will help you the most - we've got you covered today and tomorrow.

Feedback

We welcome your feedback. Email us at feedback@test-preparation.ca with your comments and suggestions. We carefully review all suggestions and often incorporate reader suggestions into upcoming versions. As a Print on Demand Publisher, we update our products frequently.

Contents

8   **Getting Started**
      The BC Police Study Plan   9
      Making a Study Schedule   11

16   **Order of Operation**
      Order Of Operation Practice   19
      Answer Key   21

23   **Exponents**
      Exponents: Tips and Short-cuts   23
      Practice Questions   26
      Answer Key   29

30   **How to Solve Word Problems**
      Types of Word Problems   33
      Word Problem Practice with Video   44
      Word Problem Practice   46
      Answer Key   54

64   **Basic Geometry**
      Pythagorean Geometry   71
      Quadrilaterals   74
      Calculating Perimeter, Area & Volume   76
      Practice Questions   78
      Answer Key   98

110   **Basic Algebra**
      One-Variable Linear Equations   110
      Two-Variable Linear Equations   111
      Simplifying Polynomials   113
      Factoring Polynomials   114
      Quadratic Equations   115
      Practice Questions   118
      Answer Key   128

149 How to Answer Basic Math Questions
        Strategy and Short-cuts   152

154 How to Study for a Math Test

158 How to Prepare for a Test
        The Strategy of Studying   160

163 How to Take a Test
        Reading the Instructions
        How to Take a Test - The Basics   166
        In the Test Room – What you MUST do! 172
        Avoid Anxiety Before a Test   178
        Common Test-Taking Mistakes   180

184 Conclusion

185 Online Resources

Getting Started

CONGRATULATIONS! By deciding to take the BC Police Entrance Test, (JIBC) you have taken the first step toward a great future! Of course, there is no point in taking this important examination unless you intend to do your best to earn the highest grade that you possibly can. That means getting yourself organized and discovering the best approaches, methods and strategies to master the material. Yes, that will require real effort and dedication on your part, but if you are willing to focus your energy and devote the study time necessary, before you know it you will be passing the BC Police Entrance Test with a great score!

We know that taking on a new endeavor can be scary, and it is easy to feel unsure of where to begin. That's where we come in. This workbook is designed to help you improve your test-taking skills, show you a few tricks of the trade and increase both your competency and confidence.

The BC Police Entrance Test Math Content

- Basic arithmetic operations

- Decimals, fractions and percent

- Equations including quadratic, linear and inequalities

- Polynomials - basic operations of polynomials including addition, subtractions, simplifying, factorizing.

- Simultaneous equation

- Basic Geometry

The BC Police Study Plan

Now that you have made the decision to take the BC Police Entrance Test, it's time to get started. Before you do another thing, you will need to figure out a plan of attack. The very best study tip is to start early! The longer the time period you devote to regular study practice, the more likely you will retain the material and be able to reach it quickly. If you thought that 1x20 is the same as 2x10, guess what? It really is not, when it comes to study time. Reviewing material for just an hour per day over the course of 20 days is far better than studying for two hours a day for only 10 days. The more often you revisit a particular piece of information, the better you will know it. Not only will your grasp and understanding be better, but your ability to reach into your brain and quickly and efficiently pull out the tidbit you need, will be greatly enhanced as well.

The great Chinese scholar and philosopher Confucius believed that true knowledge could be defined as knowing what you know and what you do not know. The first step in preparing for the BC Police

Exam is to assess your strengths and weaknesses. You may already have an idea of what you know and what you do not know, but evaluating yourself using our self-assessment modules for each of the math areas will clarify the details.

Making a Study Schedule

To make your study time most productive, you will need to develop a study plan. The purpose of the plan is to organize all the bits of pieces of information in such a way that you will not feel overwhelmed. Rome was not built in a day, and learning everything you will need to know to pass the BC Police Exam is going to take time, too. Arranging the material you need to learn into manageable chunks is the best way to go. Each study session should make you feel as though you have accomplished your goal, and your goal is simply to learn what you planned to learn during that particular session. Try to organize the content in such a way that each study session builds on previous ones. That way, you will retain the information, be better able to reach it, and review the previous bits and pieces at the same time.

The Best Study Tip! The very best study tip is to start early! The longer you study regularly, the more you will retain and 'learn' the material. Studying for 1 hour per day for 20 days is far better than studying for 2 hours for 10 days.

What don't you know?

The first step is to assess your strengths and weaknesses.

Below is a table to assess your exam readiness in each math content area.

Exam Readiness Assessment

Exam Component	Rate 1 to 5
Basic Math	
Number Operations	
Fractions, Percent and Decimals	
Exponents	
Algebra	
Solving 1 and 2 variable equations	
Solving Polynomials	
Operations with Polynomials	
Quadratics	
Inequalities	
Measurement and Geometry	
Calculate Perimeter, Circumference and Volume	
Pythagorean Theorem	
Slope of a line	

Making a Study Schedule

The key to a study plan is to divide the material you need to learn into manageable size and learn it, while at the same time reviewing the material that you already know.

Using the table above, any scores of 3 or below, you need to spend time learning, going over, and practicing this subject area. A score of 4 means you need to review the material, but you don't have to spend time re-learning. A score of 5 and you are OK with just an occasional review before the exam.

A score of 0 or 1 means you really need to work on this area and should allocate the most time and the highest priority. Some students prefer a 5-day plan and others a 10-day plan. It also depends on how much time until the exam.

Here is an example of a 5-day plan based on an example from the table above:

Exponents: 1 Study 1 hour everyday–review on last day
Fractions: 3 Study 1 hour for 2 days then ½ hour and then review
Algebra: 4 Review every second day
Quadratics: 2 Study 1 hour on the first day – then ½ hour everyday
Pythagorean Theorem: 5 Review for ½ hour every other day
Geometry: 5 Review for ½ hour every other day

Using this example, geometry and Pythagorean Theorem are good and only need occasional review. Algebra is good and needs 'some' review. Fractions need a bit of work, grammar and usage needs a lot of work and Exponents is very weak and need most time. Based on this, here is a sample study plan:

― Getting Started ―

Day	Subject	Time
Monday		
Study	Exponents	1 hour
Study	Quadratics	1 hour
	½ hour break	
Study	Fractions	1 hour
Review	Algebra	½ hour
Tuesday		
Study	Exponents	1 hour
Study	Quadratics	½ hour
	½ hour break	
Study	Fractions	½ hour
Review	Algebra	½ hour
Review	Geometry	½ hour
Wednesday		
Study	Exponents	1 hour
Study	Quadratics	½ hour
	½ hour break	
Study	Fractions	½ hour
Review	Geometry	½ hour
Thursday		
Study	Exponents	½ hour
Study	Quadratics	½ hour
Review	Fractions	½ hour
	½ hour break	
Review	Geometry	½ hour
Review	Algebra	½ hour
Friday		
Review	Exponents	½ hour
Review	Quadratics	½ hour
Review	Fractions	½ hour
	½ hour break	
Review	Algebra	½ hour
Review	Quadratics	½ hour

Using this example, adapt the study plan to your own schedule. This schedule assumes 2 ½ - 3 hours available to study everyday for a 5 day period.

First, write out what you need to study and how much. Next figure out how many days before the test. Note, do NOT study on the last day before the test. On the last day before the test, you won't learn anything and will probably only confuse yourself.

Make a table with the days before the test and the number of hours you have available to study each day. We suggest working with 1 hour and ½ hour time slots.

Start filling in the blanks, with the subjects you need to study the most getting the most time and the most regular time slots (i.e. everyday) and the subjects that you know getting the least time (e.g. ½ hour every other day, or every 3rd day).

Tips for making a schedule

Once you make a schedule, stick with it! Make your study sessions reasonable. If you make a study schedule and don't stick with it, you set yourself up for failure. Instead, schedule study sessions that are a bit shorter and set yourself up for success! Make sure your study sessions are do-able. Studying is hard work, but after you pass, you can party and take a break!

Schedule breaks. Breaks are just as important as study time. Work out a rotation of studying and breaks that works for you.

Getting Started

Build up study time. If you find it hard to sit still and study for 1 hour straight through, build up to it. Start with 20 minutes, and then take a break. Once you get used to 20-minute study sessions, increase the time to 30 minutes. Gradually work you way up to 1 hour.

40 minutes to 1 hour is optimal. Studying for longer than this is tiring and not productive. Studying for shorter isn't long enough to be productive.

Order of Operation

ORDER OF OPERATION IS AN IMPORTANT PART OF BASIC MATH.

There will probably be only a few (if any) questions specifically on Order of Operation, although they will be included in other questions.

Consider a problem like 3 + (35 - 21) x 2, which requires addition, subtraction and multiplication operations. Which of the operations to perform first? Starting with the wrong operation will give you the wrong answer. To solve this dilemma and to avoid confusion, the Order of Operation rules were set.

Order of operation is a set of mathematical rules designed to be used for calculations that require more than one arithmetic operation.

The order of operation rules are:

Rule 1: Start with calculations inside parentheses.

Rule 2: Then, solve all multiplication and division, from left to right.

Rule 3: Finally, solve all addition and subtraction, from left to right.

Order of Operation

Example 1

Solve 16 + 5 x 8

Based on the rules above, start with multiplication.

16 + (5 * 8)   Include brackets to clarify

16 + 40 = 56

Take note that if the rule was not followed and addition was done first, the answer would be different, as in:

16 + 5 x 8

21 x 8 = 168 (wrong answer)

Example 2

3 +(35 - 21) x 2

Solve the problems in parenthesis first. Then we do the multiplication, before doing the addition.

3 + (35 - 21) x 2

3 + (14) x 3

3 + 42

= 45

Answer Sheet

Order Of Operation Practice Questions

1. 7 + 2 x (6 + 3) ÷ 3 - 7 =

    a.   6
    b.   5
    c. 7
    d. 4

2. 11 + 19 x 2 =

    a. 60
    b. 50
    c. 49
    d. 54

3. (14 + 2) x 2 + 3 =

    a. 21
    b. 35
    c. 80
    d. 43

4. 120 ÷ (6 + 12 x 2)

    a. 150
    b. 40
    c. 6
    d. 4

BC Police Math Practice

5. 12 + 2 x 44

   a. 100
   b. 616
   c. 110
   d. 600

6. 10 x 2 – (7 + 9)

   a. 21
   b. 16
   c. 4
   d. 13

Answer Key

1. A

$7 + 2 \times (6 + 3) \div 3 - 7 =$

$7 + 2 \times (9) \div 3 - 7$ Brackets first

$7 + 2 \times (9) \div 3 - 7$ left to right multiplication and division

$7 + ((2 \times 9) \div 3) - 7$

$= 7 + (18 \div 3) - 7$

$= 7 + 6 - 7$ Now left to right addition and subtraction

$= 13 - 7 = 6$

2. C
$11 + 19 \times 2$ first, left to right multiplication

$11 + 38$

$= 49$

3. B
$(14 + 2) \times 2 + 3 =$ Operations inside parenthesis first

$16 \times 2 + 3$ Next left to right multiplication

$32 + 3 = 35$

4. D
$120 \div (6 + 12 \times 2)$ Operations inside parenthesis first - multiplication before addition

120 ÷ (6 + 24) Now addition in parenthesis
120 ÷ 30 = 4

5. A

12 + 2 x 44 Multiplication first

12 + 88 = 100

6. C

10 x 2 − (7 + 9) Operations inside parenthesis first

10 x 2 − 16 Next, left to right, multiplication

20 - 16 = 4

Exponents

Exponents and simplifying expressions with radicals are part of the Numbers and Operations section of the BC Police math test.

Exponents: Tips, Tricks and Short-cuts

Exponents are just shorthand for saying that you're multiplying a number by itself two or more times.

For instance, instead of saying 5 x 5 x 5, you can show that you're multiplying 5 by itself 3 times if you just write 5^3.

We usually say this as "five to the third power" or "five to the power of three." In this example, the raised 3 is an "exponent," and the 5 is the "base."

You can even use exponents with fractions. For instance, $1/2\ ^3$ means you're multiplying 1/2 x 1/2 x 1/2. (The answer is 1/8).

Multiplying Exponents

For exponents with the same base, for instance 5^3 X 5^2, add the exponents and keep the same base. The answer, then, is 5^5.
If the bases are different, for example, in 5^3 X 3^2,

you have to do the math the long way to figure it out.
5 x 5 x 5 = 125, and 3 X 3 = 9.

125 X 9 = 1125

Dividing Exponents

For exponents with the same base, subtract the exponents. In the problem above, 5^3 X 5^2, 3 - 2 = 1. 5 to the power of 1 is 5.

Here are some Quick things to remember

Any number to the power of 1 is that number.

Any number raised to the power of 0 is 1.

Number (x)	X^2	X^3
1	1	1
2	4	8
3	9	27
4	16	64
5	25	125
6	36	216
7	49	343
8	64	512
9	81	729
10	100	1000
11	121	1331
12	144	1728
13	169	2197
14	196	2744
15	225	3375

Answer Sheet

	A	B	C	D
1	○	○	○	○
2	○	○	○	○
3	○	○	○	○
4	○	○	○	○
5	○	○	○	○
6	○	○	○	○
7	○	○	○	○
8	○	○	○	○
9	○	○	○	○
10	○	○	○	○
11	○	○	○	○
12	○	○	○	○
13	○	○	○	○
14	○	○	○	○
15	○	○	○	○

Practice Questions

1. Express in 3^4 standard form

    a. 81
    b. 27
    c. 12
    d. 9

2. Simplify $4^3 + 2^4$

    a. 45
    b. 108
    c. 80
    d. 48

3. If $x = 2$ and $y = 5$, solve $xy^3 - x^3$

    a. 240
    b. 258
    c. 248
    d. 242

4. $X^3 \times X^2$

    a. 5^x
    b. x^{-5}
    c. x^{-1}
    d. X^5

— Exponents —

5. Divide 243 by 3^3

    a. 243
    b. 11
    c. 9
    d. 27

6. $7^5 - 3^5 =$

    a. 15,000
    b. 16,564
    c. 15,800
    d. 15,007

7. Solve for x if, $10^2 \times 100^2 = 1000^x$

    a. x = 2
    b. x = 3
    c. x = -2
    d. x = 0

8. Express 9 x 9 x 9 in exponential form and standard form.

    a. $9^3 = 719$
    b. $9^3 = 629$
    c. $9^3 = 729$
    d. $10^3 = 729$

9. Multiply 0.27 by 9^2

    a. 218.7
    b. 21.87
    c. 21
    d. 20.87

10. Solve $3^8/3^5$

    a. 3^3
    b. 3^5
    c. 3^6
    d. 3^4

— Exponents —

Answer Key

1. A
3 x 3 x 3 x 3 = 81

2. C
(4 x 4 x 4) + (2 x 2 x 2 x 2) = 64 + 16 = 80

3. D
$2(5)^3 - (2)^3$ = 2(125) – 8 = 250 – 8 = 242

4. D
$X^3 \times X^2 = X^{3+2} = X^5$

5. C
$243/3^3$ 3 x 3 x 3 = 27
243/27 = 9

6. B
(7 x 7 x 7 x 7 x 7) - (3 x 3 x 3 x 3 x 3) = 16,807 – 243 = 16,564

7. A
10 x 10 x 100 x 100 = 1000^x, =100 x 10,000 = 1000^x, = 1,000,000 = 1000x = x = 2

8. C
Exponential form is 9^3 and standard from is 729

9. B
0.27 (9 x 9) = 0.27 x 81 = 21.87

10. A
$3^{8-5} = 3^3$
To divide exponents with the same base, subtract the exponents.

How to Solve Word Problems

How to Solve Word Problems

Do you know what the biggest tip for solving word problems is?

Practice regularly and systematically.

Sounds simple and easy right? Yes it is, and yes it really does work.

Word problems are a way of thinking and require you to translate a real-world problem into mathematical terms.

Some math teachers say that learning how to think mathematically is the main reason for teaching word problems.

So what does that mean?

Studying word problems and math in general requires a logical and mathematical frame of mind. The only way you can get this is by practicing regularly, which means every day.

It is critical that you practice word problems every day for the 5 days before the exam as the absolute minimum.

If you practice and miss a day, you have lost the

mathematical frame of mind and the benefit of your previous practice is gone. You must start all over again.

Everything is important.

All the information given in the problem has some purpose. There is no unnecessary information! Word problems are typically around 50 words in 2 or 3 sentences.

Often, the relationships are complicated. To explain everything, every word counts.

Make sure that you use every piece of information.

7 steps to solving word problems

Step 1 – Read through the problem at least three times. The first reading should be a quick scan, and the next two readings should be done slowly to find answers to these questions:

> What does the problem ask? (Usually located at the end)

Mark all information and underline all important words or phrases.

Step 2 – Draw a picture. Use arrows, circles, lines, whatever works for you. This makes the problem real.

A favorite word problem is something like, 1 train leaves Station A travelling at 100 km/hr and another train leaves Station B travelling at 60 km/hr. ...

Draw a line, the two stations, and the two trains at either end.

BC Police Math Practice

Depending on the question, make a table with a blank portion to show information you don't know.

Step 3 – Assign a single letter to represent each unknown.

You may want to note the unknown that each letter represents so you don't get confused.

Step 4 – Translate the information into an equation.

Remember that the main problem with word problems is that they are not expressed in regular math equations. Your ability to identify correctly the variables and translate the information into an equation determines your ability to solve the problem.

Step 5 – Check the equation to see if it looks like regular equations that you are used to seeing and whether it looks sensible.

Does the equation appear to represent the information in the question? Take note that you may need to rewrite some formulas needed to solve the word problem equation.

Step 6 – Use algebra rules to solve the equation.

Simplify each side of the equation by removing parentheses and combining like terms.

Use addition or subtraction to isolate the variable term on one side of the equation. If a number crosses to the other side of the equation, the sign changes to the opposite -- for example positive to negative.

Use multiplication or division to solve for the variable. What you to once side of the equation you must do for the other.

Word Problem Multiple Choice

Where there are multiple unknowns you will need to use elimination or substitution methods to resolve all the equations.

Step 7 – Check your final answers to see if they make sense with the information given in the problem.

For example, if the word problem involves a discount, the final price should be less or if a product was taxed then the final answer has to cost more.

Types of Word Problems

Word problems can be classified into 12 types. Below are examples of each type with a complete solution. Some types of word problems can be solved quickly using multiple choice strategies and some cannot. Always look for ways to estimate the answer and then eliminate choices.

1. Age

A girl is 10 years older than her brother. By next year, she will be twice the age of her brother. What are their ages now?

    a. 25, 15
    b. 19, 9
    c. 21, 11
    d. 29, 19

Solution: B

We will assume that the girl's age is "a" and her brother's age is "b." This means that based on the information in the first sentence,

a = 10 + b
Next year, she will be twice her brother's age, which gives, a + 1 = 2(b + 1)

We need to solve for one unknown factor and then use the answer to solve for the other. To do this we substitute the value of "a" from the first equation into the second equation. This gives

10+b + 1 = 2b + 2
11 + b = 2b + 2
11 − 2 = 2b − b
b= 9

9 = b this means that her brother is 9 years old. Solving for the girl's age in the first equation gives a = 10 + 9. a = 19 the girl is aged 19. So, the girl is aged 19 and the boy is 9

2. Distance or Speed

Two boats travel down a river towards the same destination, starting at the same time. One is traveling at 52 km/hr, and the other boat at 43 km/hr. How far apart will they be after 40 minutes?

    a. 46.67 km
    b. 19.23 km
    c. 6.4 km
    d. 14.39 km

Solution: C

After 40 minutes, the first boat will have traveled = 52 km/hr x 40 minutes/60 minutes = 34.7 km
After 40 minutes, the second boat will have traveled = 43 km/hr x 40/60 minutes = 28.66 km
Difference between the two boats will be 34.7 km − 28.66 km = 6.04 km.

Multiple Choice Strategy

First estimate the answer. The first boat is traveling 9 km. faster than the second, for 40 minutes, which is 2/3 of an hour. 2/3 of 9 = 6, as a rough guess of the distance apart.

Choices A, B and D can be eliminated right away.

3. Ratio

The instructions in a cookbook state that 700 grams of flour must be mixed in 100 ml of water, and 0.90 grams of salt added. A cook however has just 325 grams of flour. What is the quantity of water and salt that he should use?

    a. 0.41 grams and 46.4 ml
    b. 0.45 grams and 49.3 ml
    c. 0.39 grams and 39.8 ml
    d. 0.25 grams and 40.1 ml

Solution: A

The Cookbook states 700 grams of flour, but the cook only has 325. The first step is to determine the percentage of flour he has 325/700 x 100 = 46.4%
That means that 46.4% of all other items must also be used.
46.4% of 100 = 46.4 ml of water
46.4% of 0.90 = 0.41 grams of salt.

Multiple Choice Strategy

The recipe calls for 700 grams of flour but the cook only has 325, which is just less than half, the quantity of water and salt are going to be about half.

Choices C and D can be eliminated right away. Choice B is very close so be careful. Looking closely at choice B, it is exactly half, and since 325 is slightly less than half of 700, it can't be correct.

Choice A is correct.

4. Percent

An agent received $6,685 as his commission for selling a property. If his commission was 13% of the selling price, how much was the property?

  a. $68,825
  b. $121,850
  c. $49,025
  d. $51,423

Solution: D

Let's assume that the property price is x. That means from the information given, 13% of x = 6,685 Solve for x,

x = 6685 x 100/13 = $51,423

Multiple Choice Strategy

The commission, 13%, is just over 10%, which is easier to work with. Round up $6685 to $6700, and multiple by 10 for an approximate answer. 10

Word Problems

X 6700 = $67,000. You can do this in your head. Choice B is much too big and can be eliminated. Choice C is too small and can be eliminated. Choices A and D are left and good possibilities.

Do the calculations to make the final choice.

5. Sales & Profit

A store owner buys merchandise for $21,045. He transports them for $3,905 and pays his staff $1,450 to stock the merchandise on his shelves. If he does not incur further costs, how much does he need to sell the items to make $5,000 profit?

    a. $32,500
    b. $29,350
    c. $32,400
    d. $31,400

Solution: D

Total cost of the items is $21,045 + $3,905 + $1,450 = $26,400

Total cost is now $26,400 + $5000 profit = $31,400

Multiple Choice Strategy

Round off and add the numbers up in your head quickly.
21,000 + 4,000 + 1500 = 26500. Add in 5000 profit for a total of 31500.

Choice B is too small and can be eliminated. Choice C and Choice A are too large and can be eliminated.

6. Tax/Income

A woman earns $42,000 per month and pays 5% tax on her monthly income. If the Government increases her monthly taxes by $1,500, what is her income after tax?

    a. $38,400
    b. $36,050
    c. $40,500
    d. $39, 500

Solution: A

Initial tax on income was 5/100 x 42,000 = $2,100
$1,500 was added to the tax to give $2,100 + 1,500 = $3,600
Income after tax is $42,000 - $3,600 = $38,400

7. Simple Interest Word Problems

Simple interest is one type of interest problems. There are always four variables of any simple interest equation. With simple interest, you would be given three of these variables and be asked to solve for one unknown variable. With more complex interest problems, you would have to solve for multiple variables.

The four variables of simple interest are:

    P – Principal which refers to the original amount of money put in the account
    I – Interest or the amount of money earned as interest
    r – Rate or interest rate. This MUST ALWAYS be in decimal format and not in percentage
    t – Time or the amount of time the money is kept in the account to earn interest

Word Problems

The formula for simple interest is I = P x r x t

Example 1

A customer deposits $1,000 in a savings account with a bank that offers 2% interest. How much interest will be earned after 4 years?

For this problem, there are 3 variables as expected.

P = $1,000
t = 4 years
r = 2%
I = ?

Before we can begin solving for I using the simple interest formula, we need to first convert the rate from percentage to decimal.

2% = 2/100 = 0.02
Now we can use the formula: I = P x r x t

I = 1,000 x 0.02 x 4 = 80
This means that the $1,000 would have earned an interest of $80 after 4 years. The total in the account after 4 years will thus be principal + interest earned, or 1,000 + 80 = $1,080

Example 2

Sandra deposits $1400 in a savings account with a bank at 5% interest. How long will she have to leave the money in the bank to earn $420 as interest to buy a second-hand car?

BC Police Math Practice

In this example, the given information is:
I = $420
P = $1,400
r - 5%
t - ?

As usual, first we convert the rate from percentage to decimal

5% = 5/100 = 0.05

Next, we plug in the variables we know into the simple interest formula - I = P x r x t

420 = 1,400 x 0.05 x t
420 = 70 x t
420 = 70t
t = 420/70
t = 6

Sandra will have to leave her $1,400 in the bank for 6 years to earn her an interest of $420 at a rate of 5%.

Other important simple interest formula to remember are below. To use these formula, do not convert r (rate) to decimal.

P = 100 x interest/ r x t
r = 100 x interest/p x t
t = 100 x interest/ p x r

Word Problems

8. Averaging

The average weight of 10 books is 54 grams. 2 more books were added and the average weight became 55.4. If one of the 2 new books added weighed 62.8 g, what is the weight of the other?

    a. 44.7 g
    b. 67.4 g
    c. 62 g
    d. 52 g

Solution: C

Total weight of 10 books with average 54 grams will be
= 1 0 × 54 = 540 g

Total weight of 12 books with average 55.4 will be
= 55.4 × 12 = 664.8 g

Total weight of the remaining 2 will be
= 664.8 – 540 = 124.8 g

If one weighs 62.8, the weight of the other will be
= 124.8 g – 62.8 g = 62 g

Multiple Choice Strategy

Averaging problems can be estimated by looking at which direction the average goes. If additional items are added and the average goes up, the new items much be greater than the average. If the average goes down after new items are added, the new items must be less than the average.

Here, the average is 54 grams and 2 books are added which increases the average to 55.4, so the new books must weight more than 54 grams.
Choices A and D can be eliminated right away.

9. Probability

A bag contains 15 marbles of various colors. If 3 marbles are white, 5 are red and the rest are black, what is the probability of randomly picking out a black marble from the bag?

 a. 7/15
 b. 3/15
 c. 1/5
 d. 4/15

Solution: A

Total marbles = 15
Number of black marbles = 15 − (3 + 5) = 7
Probability of picking out a black marble = 7/15

10. Two Variables

A company paid a total of $2850 to book for 6 single rooms and 4 double rooms in an hotel for one night. Another company paid $3185 to book for 13 single rooms for one night in the same hotel. What is the cost for single and double rooms in that hotel?

 a. single= $250 and double = $345
 b. single= $254 and double = $350
 c. single = $245 and double = $305
 d. single = $245 and double = $345

Solution: D

We can determine the price of single rooms from the information given of the second company. 13 single rooms = 3185.
One single room = 3185 / 13 = 245

The first company paid for 6 single rooms at $245.
245 x 6 = $1470

Total amount paid for 4 double rooms by first company = $2850 - $1470 = $1380

Cost per double room = 1380 / 4 = $345

11. Geometry

The length of a rectangle is 5 in. more than its width. The perimeter of the rectangle is 26 in. What is the width and length of the rectangle?

    a. width = 6 inches, Length = 9 inches
    b. width = 4 inches, Length = 9 inches
    c. width = 4 inches, Length = 5 inches
    d. width = 6 inches, Length = 11 inches

Solution: B

Formula for perimeter of a rectangle is 2(L + W)
p = 26, so 2(L + W) = p

The length is 5 inches more than the width, so
2(w + 5) + 2w = 26
2w + 10 + 2w = 26
2w + 2w = 26 - 10
4w = 16

W = 16/4 = 4 inches

L is 5 inches more than w, so L = 5 + 4 = 9 inches.

12. Totals and fractions

A basket contains 125 oranges, mangoes and apples. If 3/5 of the fruits in the basket are mangoes and only 2/5 of the mangoes are ripe, how many ripe mangoes are there in the basket?

    a. 30
    b. 68
    c. 55
    d. 47

Solution: A
Number of mangoes in the basket is 3/5 x 125 = 75
Number of ripe mangoes = 2/5 x 75 = 30

Word Problem Practice with Video Solutions

HTTPS://YOUTU.BE/6XWA6FO6YCE

Answer Sheet

	A	B	C	D	E			A	B	C	D	E
1	○	○	○	○	○		21	○	○	○	○	○
2	○	○	○	○	○		22	○	○	○	○	○
3	○	○	○	○	○		23	○	○	○	○	○
4	○	○	○	○	○		24	○	○	○	○	○
5	○	○	○	○	○		25	○	○	○	○	○
6	○	○	○	○	○							
7	○	○	○	○	○							
8	○	○	○	○	○							
9	○	○	○	○	○							
10	○	○	○	○	○							
11	○	○	○	○	○							
12	○	○	○	○	○							
13	○	○	○	○	○							
14	○	○	○	○	○							
15	○	○	○	○	○							
16	○	○	○	○	○							
17	○	○	○	○	○							
18	○	○	○	○	○							
19	○	○	○	○	○							
20	○	○	○	○	○							

Word Problem Practice

1. Translate the following into an equation: Five greater than 3 times a number.

    a. 3X + 5
    b. 5X + 3
    c. (5 + 3)X
    d. 5(3 + X)

2. Translate the following into an equation: three plus a number times 7 equals 42.

    a. 7(3 + X) = 42
    b. 3(X + 7) = 42
    c. 3X + 7 = 42
    d. (3 + 7)X = 42

3. Translate the following into an equation: 2 + a number divided by 7.

    a. (2 + X)/7
    b. (7 + X)/2
    c. (2 + 7)/X
    d. 2/(7 + X)

4. Translate the following into an equation: six times a number plus five.

    a. 6X + 5
    b. 6(X+5)
    c. 5X + 6
    d. (6 * 5) + 5

Word Problems

5. A box contains 7 black pencils and 28 blue ones. What is the ratio between the black and blue pens?

    a.   1:4
    b.   2:7
    c. 1:8
    d.   1:9

6. The manager of a weaving factory estimates that if 10 machines run at 100% efficiency for 8 hours, they will produce 1450 meters of cloth. Due to some technical problems, 4 machines run of 95% efficiency and the remaining 6 at 90% efficiency. How many meters of cloth can these machines will produce in 8 hours?

    a. 1334 meters
    b. 1310 meters
    c. 1300 meters
    d. 1285 meters

7. In a local election at polling station A, 945 voters cast their vote out of 1270 registered voters. At polling station B, 860 cast their vote out of 1050 registered voters and at station C, 1210 cast their vote out of 1440 registered voters. What is the total turnout from all three polling stations?

    a. 70%
    b. 74%
    c. 76%
    d. 80%

8. If Lynn can type a page in p minutes, what portion of the page can she do in 5 minutes?

    a. p/5
    b. p - 5
    c. p + 5
    d. 5/p

9. If Sally can paint a house in 4 hours, and John can paint the same house in 6 hours, how long will it take for both to paint a house?

    a. 2 hours and 24 minutes
    b. 3 hours and 12 minutes
    c. 3 hours and 44 minutes
    d. 4 hours and 10 minutes

10. Employees of a discount appliance store receive an additional 20% off the lowest price on any item. If an employee purchases a dishwasher during a 15% off sale, how much will he pay if the dishwasher originally cost $450?

    a. $280.90
    b. $287.00
    c. $292.50
    d. $306.00

11. The sale price of a car is $12,590, which is 20% off the original price. What is the original price?

    a. $14,310.40
    b. $14,990.90
    c. $15,108.00
    d. $15,737.50

Word Problems

12. Richard gives 's' amount of salary to each of his 'n' employees weekly. If he has 'x' amount of money, how many days he can employ these 'n' employees.

    a. sx/7n
    b. 7x/nx
    c. nx/7s
    d. 7x/ns

13. A distributor purchased 550 pounds of potatoes for $165. He distributed these at a rate of $6.4 per 20 pounds to 15 shops, $3.4 per 10 pounds to 12 shops and the remainder at $1.8 per 5 pounds. If his total distribution cost is $10, what will his profit be?

    a. $10.40
    b. $8.60
    c. $14.90
    d. $23.40

14. How much pay does Mr. Johnson receive if he gives half of his pay to his family, $250 to his landlord, and has exactly 3/7 of his pay left over?

    a. $3600
    b. $3500
    c. $2800
    d. $1750

15. The cost of waterproofing canvas is .50 a square yard. What's the total cost for waterproofing a canvas truck cover that is 15' x 24'?

    a. $18.00
    b. $6.67
    c. $180.00
    d. $20.00

16. The price of a book went from $20 to $25. What percent did the price increase?

    a. 5%
    b. 10%
    c. 20%
    d. 25%

17. In the time required to serve 43 customers, a server breaks 2 glasses and slips 5 times. The next day, the same server breaks 10 glasses. Assuming that glasses broken is proportional to customers served, how many customers did she serve?

    a. 25
    b. 43
    c. 86
    d. 215

Word Problems

18. A square lawn has an area of 62,500 square meters. What is the cost of building fence around it at a rate of $5.5 per meter?

    a. $4000
    b. $4500
    c. $5000
    d. $5500

19. Susan wants to buy a leather jacket that costs $545.00 and is on sale for 10% off. What is the approximate cost?

    a. $525
    b. $450
    c. $475
    d. $500

20. Sarah weighs 25 pounds more than Tony. If together they weigh 205 pounds, how much does Sarah weigh in kilograms? Assume 1 pound = 0.4535 kilograms.

    a. 41
    b. 48
    c. 50
    d. 52

BC Police Math Practice

21. A man buys an item for $420 and has a balance of $3000.00. How much did he have before his purchase?

    a. $2,580

    b. $3,420

    c. $2,420

    d. $342

22. The average weight of 13 students in a class of 15 (two were absent that day) is 42 kg. When the remaining two are weighed, the average became 42.7 kg. If one of the remaining students weighs 48 kg., how much does the other weigh?

    a. 44.7 kg.

    b. 45.6 kg.

    c. 46.5 kg.

    d. 47.4 kg.

23. The total expense of building a fence around a square-shaped field is $2000 at a rate of $5 per meter. What is the length of one side?

    a. 40 meters

    b. 80 meters

    c. 100 meters

    d. 320 meters

Word Problems

24. There were some oranges in a basket. By adding 8/5 of the total to the basket, the new total is 130. How many oranges were in the basket?

    a. 60
    b. 50
    c. 40
    d. 35

25. A person earns $25,000 per month and pays $9,000 income tax per year. The Government increased income tax by 0.5% per month and his monthly earning was increased $11,000. How much more income tax will he pay per month?

    a. $1260
    b. $1050
    c. $750
    d. $510

BC Police Math Practice

Answer Key

Part 1 - Equation Translation

1. A
Five greater than 3 times a number.
5 + 3 times a number.
3X + 5

2. A
Three plus a number times 7 equals 42.
Let X be the number.
(3 + X) times 7 = 42
7(3 + X) = 42

3. A
2 + a number divided by 7.
(2 + X) divided by 7.
(2 + X)/7

4. B
Six times a number plus five is the same as saying six times (a number plus five). Or,
6 * (a number plus five). Let X be the number so,
6(X + 5).

5. A
The ratio between black and blue pens is 7 to 28 or 7:28. Bring to the lowest terms by dividing both sides by 7 gives 1:4.

6. A
At 100% efficiency 1 machine produces 1450/10 = 145 m of cloth.

At 95% efficiency, 4 machines produce 4 * 145 *

95/100 = 551 m of cloth.

At 90% efficiency, 6 machines produce 6 * 145 * 90/100 = 783 m of cloth.

Total cloth produced by all 10 machines = 551 + 783 = 1334 m

Since the information provided and the question are based on 8 hours, we did not need to use time to reach the answer.

7. D
https://youtu.be/Es3Yg5pfYeY

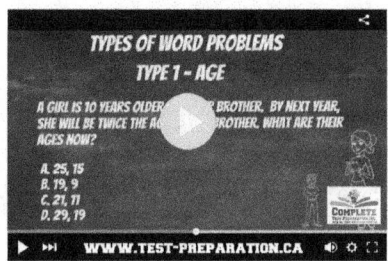

To find the total turnout in all three polling stations, we need to proportion the number of voters to the number of all registered voters.

Number of total voters = 945 + 860 + 1210 = 3015

Number of total registered voters = 1270 + 1050 + 1440 = 3760

Percentage turnout over all three polling stations = 3015 * 100/3760 = 80.19%

Checking the answers, we round 80.19 to the nearest whole number: 80%

8. D
https://youtu.be/syDAMxmkYgY

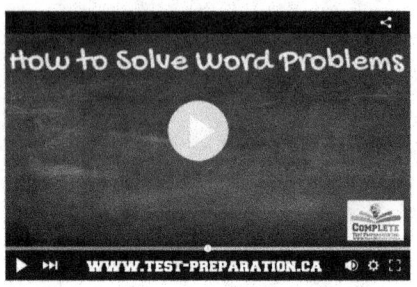

This is a simple direct proportion problem: If Lynn can type 1 page in p minutes, she can type x pages in 5 minutes

We do cross multiplication: x * p = 5 * 1

Then,

x = 5/p

9. A
This is an inverse ratio problem.

1/x = 1/a + 1/b where a is the time Sally can paint a house, b is the time John can paint a house, x is the time Sally and John can together paint a house.

So,

1/x = 1/4 + 1/6 ... We use the least common multiple in the denominator that is 24:

1/x = 6/24 + 4/24

1/x = 10/24

x = 24/10

x = 2.4 hours.

In other words; 2 hours + 0.4 hours = 2 hours + 0.4•60 minutes

= 2 hours 24 minutes

10. D
https://youtu.be/I_FaJPJzepE

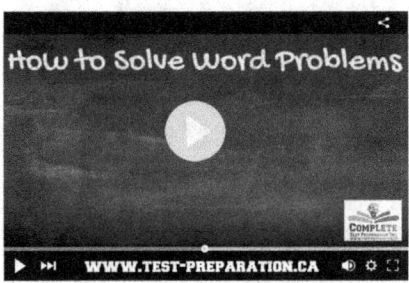

The cost of the dishwasher = $450

15% discount amount = 450•15/100 = $67.5

The discounted price = 450 – 67.5 = $382.5

20% additional discount amount on lowest price = 382.5•20/100 = $76.5

So, the final discounted price = 382.5 - 76.5 = $306.00

11. D
Original price = x,
80/100 = 12590/X,
80X = 1259000,
X = 15,737.50.

12. D
https://youtu.be/EF92e6V4mAA

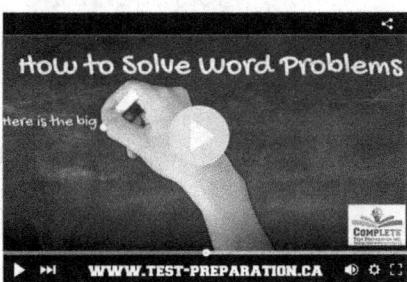

We are given that each of the n employees

earns s amount of salary weekly. This means that one employee earns s salary weekly. So; Richard has 'ns' amount of money to employ n employees for a week.

We are asked to find the number of days n employees can be employed with x amount of money. We can do simple direct proportion:

If Richard can employ n employees for 7 days with 'ns' amount of money,

Richard can employ n employees for y days with x amount of money ... y is the number of days we need to find.

We can cross multiply:

y = (x * 7)/(ns)

y = 7x/ns

13. B
The distribution is done at three different rates and in three different amounts:

$6.4 per 20 kilograms to 15 shops ... 20•15 = 300 kilograms distributed

$3.4 per 10 kilograms to 12 shops ... 10•12 = 120 kilograms distributed

550 - (300 + 120) = 550 - 420 = 130 kilograms left. This amount is distributed in 5 kilogram portions. So, this means that there are 130/5 = 26 shops.

$1.8 per 130 kilograms.

We need to find the amount he earned overall these distributions.

$6.4 per 20 kilograms : 6.4 * 15 = $96 for 300 kilograms

$3.4 per 10 kilograms : 3.4 *12 = $40.8 for 120 kilograms

$1.8 per 5 kilograms : 1.8 * 26 = $46.8 for 130 kilograms

So, he earned 96 + 40.8 + 46.8 = $ 183.6

The total distribution cost is given as $10

The profit is found by: Money earned - money spent ... It is important to remember that he bought 550 kilograms of potatoes for $165 at the beginning:

Profit = 183.6 - 10 - 165 = $8.6

14. B
We check the fractions taking place in the question. We see that there is a "half" (that is 1/2) and 3/7. So, we multiply the denominators of these fractions to decide how to name the total money. We say that Mr. Johnson has 14x at the beginning; he gives half of this, meaning 7x, to his family. $250 to his landlord. He has 3/7 of his money left. 3/7 of 14x is equal to:

14x * (3/7) = 6x

So,

Spent money is: 7x + 250

Unspent money is: 6x

Total money is: 14x

Write an equation: total money = spent money + unspent money

14x = 7x + 250 + 6x

14x - 7x - 6x = 250

x = 250

We are asked to find the total money that is 14x:

14x = 14 * 250 = $3500

15. D
First calculate total square feet, which is 15•24 = 360 ft2. Next, convert this value to square yards, (1 yards2 = 9 ft2) which is 360/9 = 40 yards2. At $0.50 per square yard, the total cost is 40 * 0.50 = $20.

16. D
Price increased by $5 ($25-$20). To calculate the percent increase:
5/20 = X/100
500 = 20X
X = 500/20
X = 25%

17. D
2 glasses are broken for 43 customers so 1 glass breaks for every 43/2 customers served, therefore 10 glasses implies (43/2) * 10 = 215 customers.

18. D
As the lawn is square, the length of one side will be the square root of the area. $\sqrt{62{,}500}$ = 250 meters. So, the perimeter is found by 4 times the length of the side of the square:

250 * 4 = 1000 meters.

Since each meter costs $5.5, the total cost of the fence will be 1000 * 5.5 = $5,500.

Word Problems

19. D

The question asks for approximate cost, so work with round numbers. The jacket costs $545.00 so we can round up to $550. 10% of $550 is 55. We can round down to $50, which is easier to work with. $550 - $50 is $500. The jacket will cost about $500.

The actual cost will be 10% X 545 = $54.50
545 – 54.50 = $490.50

20. D

Let us denote Sarah's weight by "x." Then, since she weighs 25 pounds more than Tony, so he will be x - 25. They together weigh 205 pounds which means that the sum of the two representations will be equal to 205:

Sarah : x

Tony : x - 25

x + (x - 25) = 205 ... by arranging this equation we have:

x + x - 25 = 205

2x - 25 = 205 ... we add 25 to each side to have x term alone:

2x - 25 + 25 = 205 + 25

2x = 230

x = 230/2

x = 115 pounds → Sarah weighs 115 pounds. Since 1 pound is 0.4535 kilograms, we need to multiply 115 by 0.4535 to have her weight in kilograms:

x = 115 * 0.4535 = 52.1525 kilograms → this is

equal to 52 when rounded to the nearest whole number.

21. B
(Amount Spent) $420 + $3000 (Balance) = $3,420.00

22. C
Total weight of 13 students with average 42 will be = 42 * 13 = 546 kg.

The total weight of the remaining 2 will be found by subtracting the total weight of 13 students from the total weight of 15 students: 640.5 - 546 = 94.5 kg.

94.5 = the total weight of two students. One of these students weigh 48 kg, so;

The weight of the other will be = 94.5 – 48 = 46.5 kg

23. C
Total expense is $2000 and we are informed that $5 is spent per meter. Combining these two information, we know that the total length of the fence is 2000/5 = 400 meters.

The fence is built around a square field. If one side of the square is "a," the perimeter of the square is "4a." Here, the perimeter is equal to 400 meters. So,

400 = 4a

100 = a → this means that one side of the square is equal to 100 meters

24. B
Let the number of oranges in the basket before additions = x
Then: X + 8x/5 = 130
5x + 8x = 650

650 = 13x
X = 50

25. D
The income tax per year is $9,000. So, the income tax per month is 9,000/12 = $750.

This person earns $25,000 per month and pays $750 income tax. We need to find the rate of the income tax:

Tax rate: 750 * 100/25,000 = 3%
Government increased this rate by 0.5% so it became 3.5%.

The income of the person per month is increased $11,000 so it became:

$25,000 + $11,000 = $36,000.

The new monthly income tax is: 36,000 * 3.5/100 = $1260.

Amount of increase in tax per month is:
$1260 - $750 = $510.

Basic Geometry

THE BASIC GEOMETRY IS INCLUDED IN PATTERNS, RELATIONSHIPS, AND ALGEBRA, AND THE GEOMETRY AND MEASUREMENT SECTION.

Basic Geometry includes:

- slope of a line
- Identify linear equations from a graph
- Calculate perimeter, circumference and volume
- Solve problems using the Pythagorean theorem
- Solve real world problems using the properties of geometric shapes

Cartesian Plane, Coordinate Grid and Plane

To locate points and draw lines and curves, we use the coordinate plane. It also called Cartesian coordinate plane. It is a two-dimensional surface with a coordinate grid in it, which helps us to count the units. For the counting of those units, we use x-axis (horizontal scale) and y-axis (vertical scale).

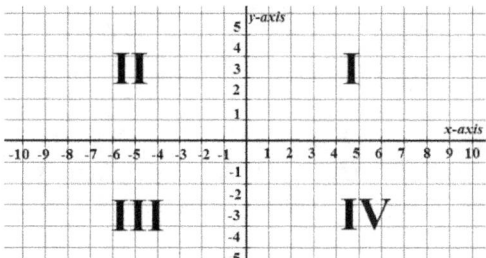

The whole system is called a coordinate system which is divided into 4 parts, called quadrants. The quadrant where all numbers are positive is the 1st quadrant (I), and if we go counterclockwise, we mark all 4 quadrants.

The location of a dot in the coordinate system is represented by coordinates. Coordinates are represented as a pair of numbers, where the 1st number is located on the x-axis and the 2nd number is located on the y-axis. So, if a dot A has coordinates a and b, then we write:

A=(a,b)   or   A(a,b)

The point where x-axis and y-axis intersect is called an origin. The origin is the point from which we measure the distance along the x and y axes.

In the Cartesian coordinate system we can calculate the distance between 2 given points. If we have dots with coordinates:
A=(a,b)
B=(c,d)

Then the distance d between A and B can be calculated by the following formula:

BC Police Math Practice

$$d = \sqrt{(c-a)^2 + (d-b)^2}$$

Cartesian coordinate system is used for the drawing of 2-dimentional shapes, and is also commonly used for functions.

Example:

Draw the function $y = (1 - x)/2$

To draw a linear function, we need at least 2 points. If we put that x=0 then value for y would be:

$$y = \frac{1-x}{2} = \frac{1-0}{2} = \frac{1}{2}$$

We found the 1st point, let's name it A, with following coordinates:

A = (0,1/2)

To find the 2nd point, we can put that x=1. In this case, the value for y would be:

$$y = \frac{1-x}{2} = \frac{1-1}{2} = \frac{0}{2} = 0$$

If we denote the 2nd point with B, then the coordinates for this point are:

B=(1,0)

Since we have 2 points necessary for the function,

we find them in the coordinate system and we connect them with a line that represents the function,

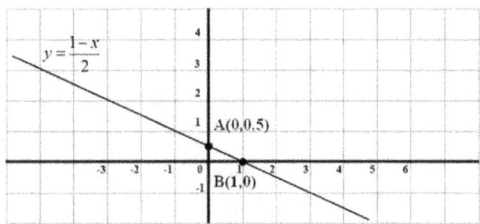

Perimeter Area and Volume

Perimeter and Area (2-dimentional shapes)

Perimeter of a shape determines the length around that shape, while the area includes the space inside the shape.

Rectangle:

$P = 2a + 2b$
$A = ab$

Square

$P = 4a$
$A = a^2$

Parallelogram

$P = 2a + 2b$
$A = ah_a = bh_b$

Rhombus

$P = 4a$
$A = ah = \dfrac{d_1 d_2}{2}$

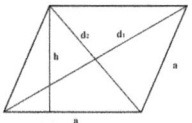

Triangle

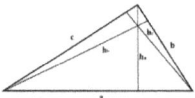
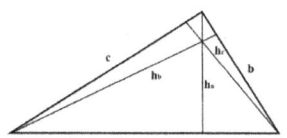

Equilateral Triangle

$P = 3a$
$A = \dfrac{a^2 \sqrt{3}}{4}$

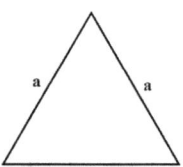

Trapezoid

$P = a + b + c + d$
$A = \dfrac{a+b}{2} h$

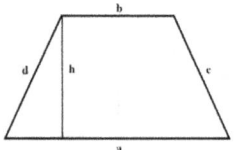

Basic Geometry

Circle

$P = 2r\pi$

$A = r^2\pi$

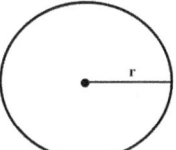

Area and Volume (3-dimentional shapes)

To calculate the area of a 3-dimentional shape, we calculate the areas of all sides and then we add them all.

To find the volume of a 3-dimentional shape, we multiply the area of the base (B) and the height (H) of the 3-dimentional shape.

$$V = BH$$

In case of a pyramid and a cone, the volume would be divided by 3.

$$V = BH/3$$

Here are some of the 3-dimentional shapes with formulas for their area and volume:

Cuboids

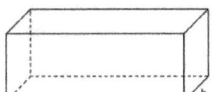 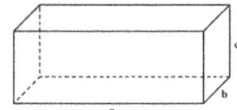

Cube

$A = 6a^2$

$V = a^3$

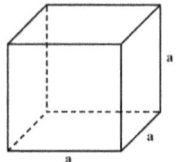

Pyramid

$A = ab + ah_a + bh_b$

$V = \dfrac{abH}{3}$

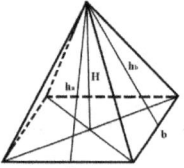

Cylinder

$A = 2r^2\pi + 2r\pi H$

$V = r^2\pi H$

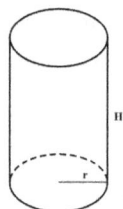

Cone

$A = (r+s)r\pi$

$V = \dfrac{r^2\pi H}{3}$

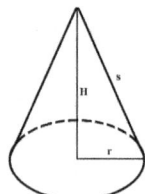

Pythagorean Geometry

If we have a right triangle ABC, where its sides (legs) are a and b and c is a hypotenuse (the side opposite the right angle), then we can establish a relationship between these sides using the following formula:

$$c^2 = a^2 + b^2$$

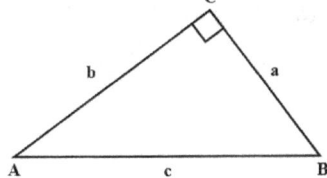

This formula is proven in the Pythagorean Theorem. There are many proofs of this theorem, but we'll look at just one geometrical proof:

If we draw squares on the right triangle's sides, then the area of the square upon the hypotenuse is equal to the sum of the areas of the squares that are upon other two sides of the triangle. Since the areas of these squares are a^2, b^2 and c^2, that is how we got the formula above.

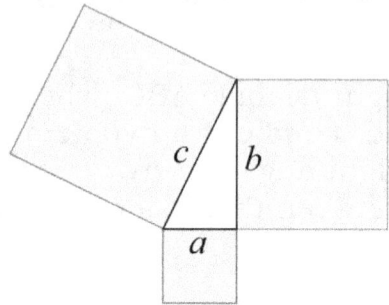

One of the famous right triangles is one with sides 3, 4 and 5. And we can see here that:

$3^2 + 4^2 = 5^2$
$9 + 16 = 25$
$25 = 25$

Example Problem:

The isosceles triangle ABC has a perimeter of 18 centimeters, and the difference between its base and legs is 3 centimeters. Find the height of this triangle.

We write the information we have about triangle ABC and we draw a picture for a better understand-

$$P = 4a$$

$$A = a^2$$

ing of the relation between its elements:

Basic Geometry

P = 18 cm
a - b = 3 cm
h = ?
We use the formula for the perimeter of the isosceles triangle, since that is what is given to us:
P = a + 2b = 18 cm

Notice that we have 2 equations with 2 variables, so we can solve it as a system of equations:

a + 2b = 18
a − b = 3 / a + 2b = 18
2a − 2b = 6 / a + 2b + 2a − 2b = 18 + 6
3a = 24
a = 24/3 = 8 cm

Now we go back to find b:
a - b = 3
8 - b = 3
b = 8 - 3
b = 5 cm

Using Pythagorean Theorem, we can find the height using a and b, because the height falls on the side a at the right angle. Notice that height cuts side Here in half, and that's why we use in the formula a/2. Here, b is the hypotenuse, so we have:

$b^2 = (a/2)^2 + h^2$
$h^2 = b^2 - (a/2)^2$
$h^2 = 5^2 - (8/2)^2$
$h^2 = 5^2 - (8/2)^2$
$h^2 = 25 - 4^2$
$h^2 = 26 - 16$
$h^2 = 9$
h = 3 cm.

Quadrilaterals

Quadrilaterals are 2-dimentional geometrical shapes that have 4 sides and 4 angles. There are many types of quadrilaterals, depending on the length of its sides, if they are parallel, and the size of its angles. All quadrilaterals have the following properties:

Sum of all interior angles is $360°$

Sum of all exterior angles is $360°$

A quadrilateral is a parallelogram is it fulfills at least one of the following conditions:

Angles on each side are supplementary
Opposite angles are equal
Opposite sides are equal
Diagonals intersect each other exactly in half

Here are some of the quadrilaterals:

Square

All sides are equal
All angles are right angles

Rectangle

2 pairs of equal sides
All angles are right angles

Parallelogram

2 pairs of equal sides
Opposite angles are equal

Rhombus

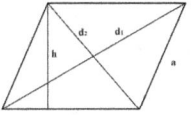

All sides are equal
Opposite angles are equal

Trapezoid

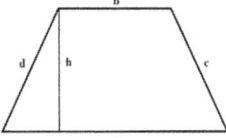

One pair of parallel sides

Example Problem

Find all angles of a parallelogram if one angle is greater than the other one by $40°$.

First, we draw an image of a parallelogram:

We denote angles by α and β, Since this is a parallelogram, the opposite angles are equal.

We are given that one angle is greater than the other one by $40°$, so we can write:

$β = α + 40°$

We solve this problem in two ways:
1) The sum of all internal angles of every quadrilateral is $360°$. There are 2 α and 2 β. So we have:

$2α + 2β = 360°$
Now, instead of β we write α + 40:
$2α + 2(α + 40°) = 360°$
$2α + 2α + 80° = 360°$
$4α = 360° - 80°$
$4α = 280°$
$α = 280° / 4$
$α = 70°$
Now we can find β from α:
$β = α + 40°$
$β = 70° + 40°$
$β = 110°$

2) One condition for parallelogram is " Angles on each side are supplementary" and we can use that to find these angles:
$α + β = 180°$
$α + α + 40° = 180°$
$2α = 180° - 40°$
$2α = 140°$
$α = 70°$

Now we find β:
$β = α + 40°$
$β = 70° + 40°$
$β = 110°$

Calculating Perimeter, Area and Volume - Quick Reference

	Circle	**Triangle**	**Square**	**Rectangle**
Perimeter	$2πr$	$a + b + c$	$4a$	$2(H + w)$
Area	$πr^2$	$1/2 bh$	$2a$	lw
Volume	$4/3 πr^3$ (Sphere)	$1/3 bh$ (pyramid)	a^3 (cube)	hwl or ab

Basic Geometry

Answer Sheet

1. (A) (B) (C) (D)   21. (A) (B) (C) (D)
2. (A) (B) (C) (D)   22. (A) (B) (C) (D)
3. (A) (B) (C) (D)   23. (A) (B) (C) (D)
4. (A) (B) (C) (D)   24. (A) (B) (C) (D)
5. (A) (B) (C) (D)   25. (A) (B) (C) (D)
6. (A) (B) (C) (D)   26. (A) (B) (C) (D)
7. (A) (B) (C) (D)   27. (A) (B) (C) (D)
8. (A) (B) (C) (D)   28. (A) (B) (C) (D)
9. (A) (B) (C) (D)   29. (A) (B) (C) (D)
10. (A) (B) (C) (D)   30. (A) (B) (C) (D)
11. (A) (B) (C) (D)   31. (A) (B) (C) (D)
12. (A) (B) (C) (D)   32. (A) (B) (C) (D)
13. (A) (B) (C) (D)   33. (A) (B) (C) (D)
14. (A) (B) (C) (D)   34. (A) (B) (C) (D)
15. (A) (B) (C) (D)   35. (A) (B) (C) (D)
16. (A) (B) (C) (D)   36. (A) (B) (C) (D)
17. (A) (B) (C) (D)   37. (A) (B) (C) (D)
18. (A) (B) (C) (D)   38. (A) (B) (C) (D)
19. (A) (B) (C) (D)   39. (A) (B) (C) (D)
20. (A) (B) (C) (D)   40. (A) (B) (C) (D)

———— BC Police Math Practice ————

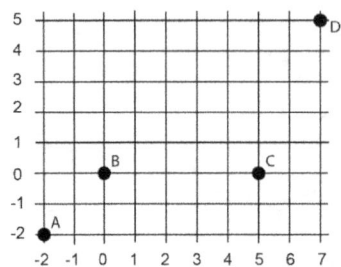

1. Which of the above points represents the origin?

    a. A
    b. B
    c. C
    d. D

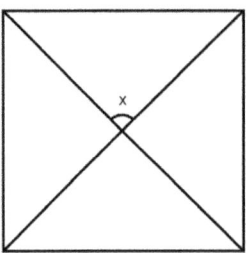

2. What is measurement of the indicated angle?

    a. 45°
    b. 90°
    c. 60°
    d. 30°

Basic Geometry

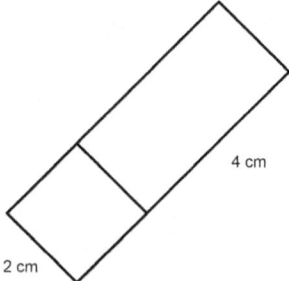

Note: figure not drawn to scale.

3. Assuming the figure with side 2 cm. is square, what is the perimeter of the above shape?

    a. 12 cm
    b. 16 cm
    c. 6 cm
    d. 20 cm

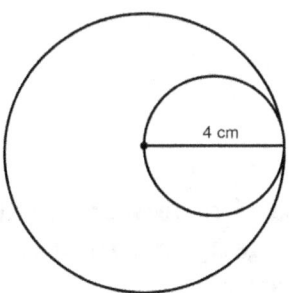

Note: Figure not drawn to scale

BC Police Math Practice

4. Assuming the diameter of the small circle is the radius of the larger circle, what is (area of large circle) - (area of small circle) in the figure above?

    a. 8π cm²
    b. 10π cm²
    c. 12π cm²
    d. 16π cm²

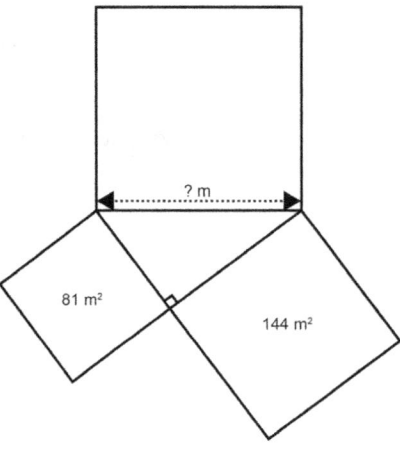

Note: Figure not drawn to scale

5. Assuming the shapes around the center right triangle are square, what is the length of each side of the indicated square above?

    a. 10
    b. 15
    c. 20
    d. 5

6. Choose the expression the figure represents.

    a.   X ≤ 1
    b.   X < 1
    c. X > 1
    d.   X ≥ 1

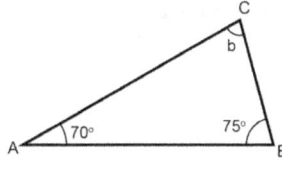

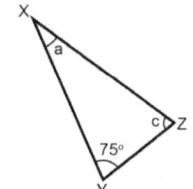

7. What are the respective values of a, b & c if both triangles are similar?

    a. 70°, 70°, 35°
    b. 70°, 35°, 70°
    c. 35°, 35°, 35°
    d. 70°, 35°, 35°

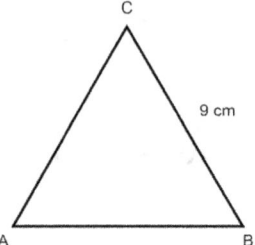

Note: figure not drawn to scale

8. What is the perimeter of the equilateral △ABC above?

    a. 18 cm
    b. 12 cm
    c. 27 cm
    d. 15 cm

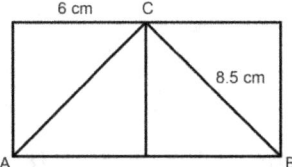

Note: figure not drawn to scale

9. **Assuming the 2 quadrangles are identical rectangles, what is perimeter of △ABC in the above shape?**

    a. 25.5 cm
    b. 27 cm
    c. 30 cm
    d. 29 cm

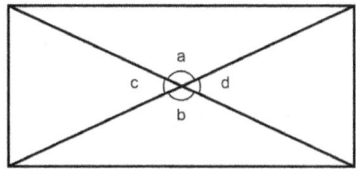

10. **What is the sum of all the angles in the rectangle above?**

    a. 180°
    b. 360°
    c. 90°
    d. 120°

BC Police Math Practice

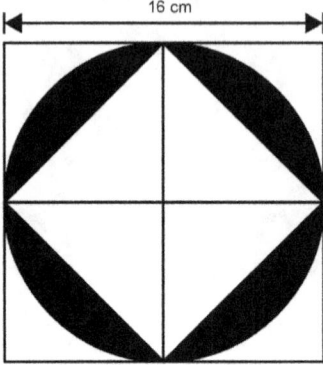

Note: figure not drawn to scale

11. A tile factory makes custom tiles, shown above, from two types of stone. If a customer requires 200 tiles, how much black stone will be required?

    a. 256 m²
    b. 2560 m²
    c. 2.56 m²
    d. 25.6 m²

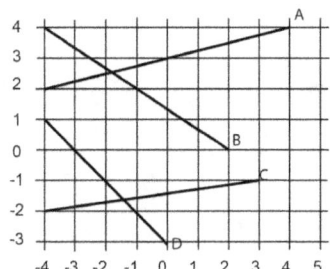

12. Which of the lines above represents the equation 2y − x = 4?

    a. A
    b. B
    c. C
    d. D

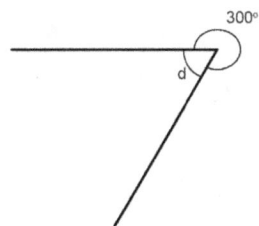

13. What is the measurement of the indicated angle?

    a. 45°
    b. 90°
    c. 60°
    d. 50°

BC Police Math Practice

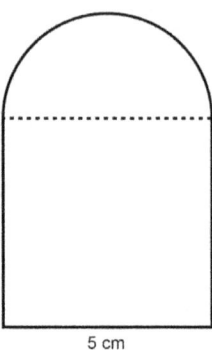

5 cm

Note: figure not drawn to scale

14. What is the perimeter of the above shape?

    a. 22.85 cm
    b. 20 cm
    c. 15 cm
    d. 25.546 cm

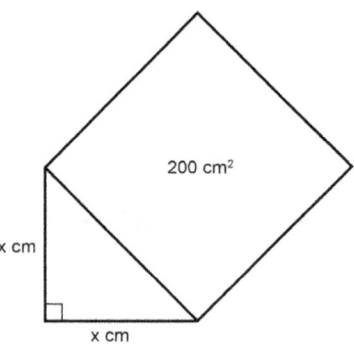

Note: Figure not drawn to scale

Basic Geometry

15. Assuming the quadrangle in the figure above is square, what is the length of the sides in the triangle above?

    a. 10
    b. 20
    c. 100
    d. 40

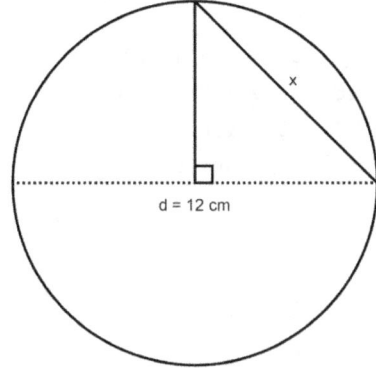

Note: Figure not drawn to scale

16. Calculate the length of side x.

    a. 6.46
    b. 8.48
    c. 3.6
    d. 6.4

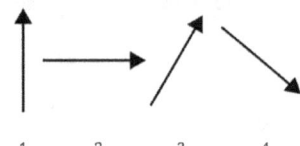

1 2 3 4

17. What is the correct order of respective slopes for the lines above?

    a. Positive, undefined, negative, positive
    b. Negative, zero, undefined, positive
    c. Undefined, zero, positive, negative
    d. Zero, positive undefined, negative

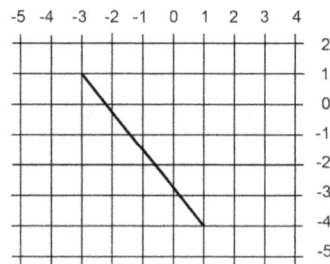

18. What is the slope of the line shown above?

    a. 5/4
    b. -4/5
    c. -5/4
    d. -4/5

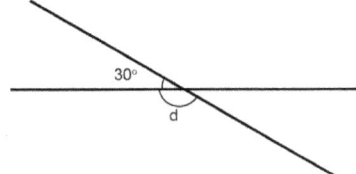

19. What is the indicated angle above?

    a. 150°
    b. 330°
    c. 60°
    d. 120°

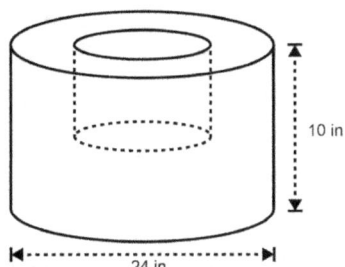

Note: figure not drawn to scale

20. What is the volume of the above solid made by a hollow cylinder that is half the size (in all dimensions) of the larger cylinder?

    a. 1440 π in³
    b. 1260 π in³
    c. 1040 π in³
    d. 960 π in³

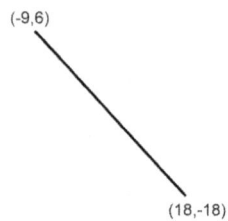

21. What is the slope of the line above?

    a.   -8/9
    b.   9/8
    c. -9/8
    d.   8/9

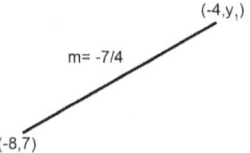

22. With the data given above, what is the value of y1?

    a.   0
    b.   -7
    c. 7
    d.   8

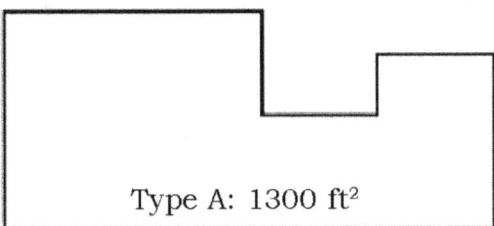

Type A: 1300 ft²

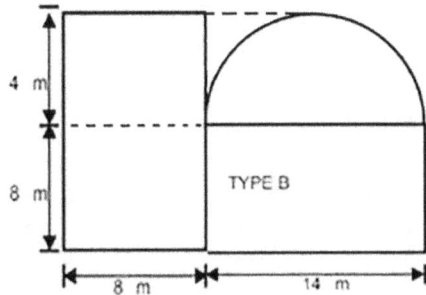

TYPE B

Note: Figure not drawn to scale

23. The price of houses in a certain subdivision is based on the total area. Susan is watching her budget and wants to choose the house with the lowest area. Which house type, A (1300 ft2) or B, should she choose if she would like the house with the lowest price? (1 m² = 10.76 ft² & π = 22/7)

 a. Type B is smaller at 140 ft²
 b. Type A is smaller
 c. Type B is smaller at 855 ft²
 d. Type B is larger

24. How much water can be stored in a cylindrical container 5 meters in diameter and 12 meters high?

Note: figure not drawn to scale

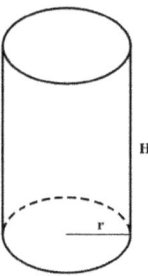

a. 235.65 m³
b. 223.65 m³
c. 240.65 m³
d. 252.65 m³

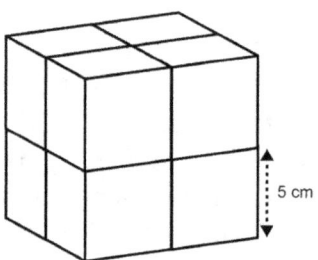

Note: figure not drawn to scale

25. Assuming the figure above is composed of cubes, what is the volume?

    a. 125 cm³

    b. 875 cm³

    c. 1000 cm³

    d. 500 cm³

26. Choose the expression the figure represents.

    a. $X > 2$

    b. $X \geq 2$

    c. $X < 2$

    d. $X \leq 2$

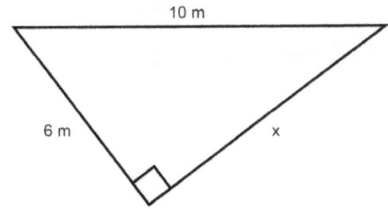

Note: figure not drawn to scale

27. What is the length of the missing side in the triangle above?

    a. 6

    b. 4

    c. 8

    d. 5

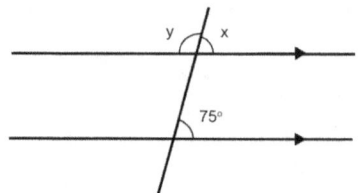

28. What is the value of the angle y?

    a. 25°
    b. 15°
    c. 30°
    d. 105°

29. What is the distance between the two points?

    a. ≈19
    b. 20
    c. ≈21
    d. ≈22

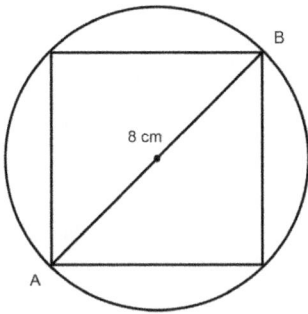

Note: figure not drawn to scale

30. What is area of the circle?

    a. 4 π cm²
    b. 12 π cm²
    c. 10 π cm²
    d. 16 π cm²

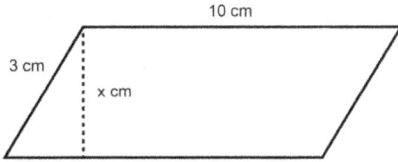

Note: figure not drawn to scale

31. What is the perimeter of the parallelogram above?

    a. 12 cm
    b. 26 cm
    c. 13 cm
    d. (13+x) cm

BC Police Math Practice

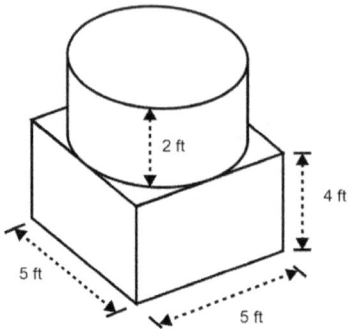

Note: figure not drawn to scale

32. What is the approximate total volume of the above solid?

    a. 120 ft³
    b. 100 ft³
    c. 140 ft³
    d. 160 ft³

33. What is the slope of the line above?

    a. 1
    b. 2
    c. 3
    d. -2

Basic Geometry

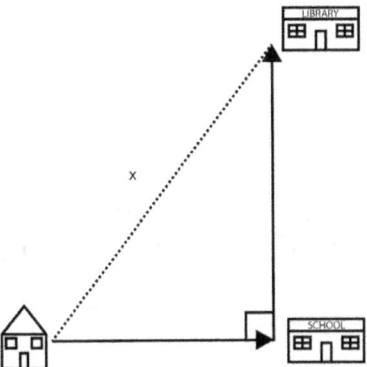

Note: figure not drawn to scale

34. Every day starting from his home Peter travels due east 3 kilometers to the school. After school he travels due north 4 kilometers to the library. What is the distance between Peter's home and the library?

    a. 15 km
    b. 10 km
    c. 5 km
    d. 12 ½ km

BC Police Math Practice

Answer Key

1. A

Point A represents the origin.

2. A

The diagonals of a square intersect at right angles, so each angle measures 90°. Half of that angle will be 45°

3. B

We see that there is a square with side 2 cm and a rectangle adjacent to it, with one side 2 cm (common side with the square) and the other side 4 cm. The perimeter of a shape is found by summing up all sides surrounding the shape, not adding the ones inside the shape. Three 2 cm sides from the square, and two 4 cm sides and one 2 cm side from the rectangle contribute the perimeter.

So, the perimeter of the shape is: 2 + 2 + 2 + 4 + 2 + 4 = 16 cm.

4. C

We are given a large circle and a small circle inside it; with the diameter equal to the radius of the large one. The diameter of the small circle is 4 cm. This means that its radius is 2 cm. Since the diameter of the small circle is the radius of the large circle, the radius of the large circle is 4 cm. The area of a circle is calculated by: πr^2 where r is the radius.

Area of the small circle: $\pi(2)^2 = 4\pi$

Area of the large circle: $\pi(4)^2 = 16\pi$

The difference area is found by:

Area of the large circle - Area of the small circle = $16\pi - 4\pi = 12\pi$

— Basic Geometry —

5. B
We see that there are three squares forming a right triangle in the middle. Two of the squares have the areas 81 m² and 144 m². If we denote their sides a and b respectively:

$a^2 = 81$ and $b^2 = 144$. The length, which is asked, is the hypotenuse; a and b are the opposite and adjacent sides of the right angle. By using the Pythagorean Theorem, we can find the value of the asked side:

Pythagorean Theorem:

(Hypotenuse)² = (Opposite Side)² + (Adjacent Side)²

$h^2 = a^2 + b^2$

$a^2 = 81$ and $b^2 = 144$ are given. So;

$h^2 = 81 + 144$

$h^2 = 225$

$h = 15$ m

6. B
The line is pointing towards numbers less than 1. The equation is therefore, X < 1.

7. D
Comparing respective angles - 70°, 35°, 35°

8. C
Equilateral triangle with 9 cm. sides
Perimeter = 9 + 9 + 9 = 27 cm.

9. D
Perimeter of triangle ABC is asked.
Perimeter of a triangle = sum of all three sides.

Here, Perimeter of △ABC = |AC| + |CB| + |AB|.

Since the triangle is located in the middle of two adjacent and identical rectangles, we find the side lengths using these rectangles:

|AB| = 6 + 6 = 12 cm

|CB| = 8.5 cm

|AC| = |CB| = 8.5 cm

Perimeter = |AC| + |CB| + |AB| = 8.5 + 8.5 + 12 = 29 cm

10. B

a + b + c + d = ?
The sum of angles around a point is 360°
a + b + c + d = 360°

11. A

Black stone for 200 tiles = 200 x [Total tile area − Inner white area(4 triangles)]

= 200 x [(16^2) - (4 x 1/2 x 8 x 8)]

= 200 x (256 - 128) = 200 x 128 = 25600 cm^2

Converting to meters − 1 cm. = 0.01 meters

= 25600/100 m^2

= 256 m^2

12. A

If a line represents an equation, all points on that line should satisfy the equation. Meaning that all (x, y) pairs present on the line should be able to verify that 2y - x is equal to 4. We can find out the correct line by trying a (x, y) point existing on each line. It is easier to choose points on the intersection of the grid lines:

Let us try the point (4, 4) on line A:

2 * 4 - 4 = 4

8 - 4 = 4

4 = 4 ... this is a correct result, so the equation for line A is 2y - x = 4.

Let us try other points to check the other lines:

Point (-1, 2) on line B:

2 * 2 - (-1) = 4

4 + 1 = 4

5 = 4 ... this is a wrong result, so the equation for line B is not 2y - x = 4.

Point (3, -1) on line C:

2 * (-1) - 3 = 4

-2 - 3 = 4

-5 = 4 ... this is a wrong result, so the equation for line C is not 2y - x = 4.

Point (-2, -1) on line D:

2 * (-1) - (-2) = 4

-2 + 2 = 4

0 = 4 ... this is a wrong result, so the equation for line D is not 2y - x = 4.

13. C
The sum of angles around a point is 360°

d + 300 = 360°

d = 60°

14. A
Find the perimeter of a shape made by merging a square and a semi circle. Perimeter = 3 sides of the square + 1/2 circumference of the circle.
= (3 × 5) + 1/2 (5 π)

= 15 + 2.5 π

= 15 + 7.853975

Perimeter = 22.85 cm

15. A
If we call one side of the square "a," the area of the square will be a^2.

We know that a^2 = 200 cm².

On the other hand; there is an isosceles right triangle. Using the **Pythagorean Theorem:**

(Hypotenuse)² = (Adjacent Side)² + (Opposite Side)² Where the hypotenuse is equal to one side of the square. So,

$a^2 = x^2 + x^2$

$200 = 2x^2$

$200/2 = 2x^2/2$

$100 = x^2$

$x = \sqrt{100}$

x = 10 cm

16. B
In the question, we have a right triangle formed inside the circle. We are asked to find the length of the hypotenuse of this triangle. We can find the other two sides of the triangle by using circle properties:

Basic Geometry

The diameter of the circle is equal to 12 cm. The legs of the right triangle are the radii of the circle; so they are 6 cm long.

Using the Pythagorean Theorem:

(Hypotenuse)² = (Adjacent Side)² + (Opposite Side)²

$x^2 = r^2 + r^2$

$x^2 = 6^2 + 6^2$

$x^2 = 72$

$x = \sqrt{72}$

$x = 8.48$

17. C
Undefined, zero, positive, negative.

18. C
Slope (m) = change in y / change in x

$(x_1, y_1) = (-3, 1)$ & $(x_2, y_2) = (1, -4)$
Slope = [-4 - 1]/[1-(-3)] = -5/4

19. A
The angles opposite both angles 30° and angle d are respectively equal to vertical angles.

2(30° + d) = 360°

2d = 360° - 60°

2d = 300°

d = 150°

20. B
Total Volume = Volume of large cylinder - Volume of small cylinder

Volume of a cylinder = area of base • height = $\pi r^2 \cdot h$

Total Volume = $(\pi * 12^2 * 10) - (\pi * 6^2 * 5)$ = $1440\pi - 180\pi$

= 1260π in³

21. A
If we know the coordinates of two points on a line, we can find the slope (m) with the below formula:

$m = (y_2 - y_1)/(x_2 - x_1)$ where (x_1, y_1) represent the coordinates of one point and (x_2, y_2) the other.

In this question:

$(-9, 6) : x_1 = -9, y_1 = 6$

$(18, -18) : x_2 = 18, y_2 = -18$

Inserting these values into the formula:

$m = (-18 - 6)/(18 - (-9)) = (-24)/(27)$ ... Simplifying by 3:

$m = -8/9$

22. A
If we know the coordinates of two points on a line, we can find the slope (m) with the below formula:
$m = (y_2 - y_1)/(x_2 - x_1)$ where (x_1, y_1) represent the coordinates of one point and (x_2, y_2) the other.

In this question:

$(-4, y_1) : x_1 = -4, y_1 =$ we will find

$(-8, 7) : x_2 = -8, y_2 = 7$

$m = -7/4$

Inserting these values into the formula:

$-7/4 = (7 - y_1)/(-8 - (-4))$

$-7/4 = (7 - y_1)/(-8 + 4)$

$7/(-4) = (7 - y_1)/(-4)$ ... Simplifying the denominators of both sides by -4:

$7 = 7 - y_1$

$0 = -y_1$

$y_1 = 0$

23. D

Area of Type B consists of two rectangles and a half circle. We can find these three areas and sum them up to find the total area:

Area of the left rectangle: $(4 + 8) * 8 = 96$ m²

Area of the right rectangle: $14 * 8 = 112$ m²

The diameter of the circle is equal to 14 m. So, the radius is $14/2 = 7$:

Area of the half circle = $(1/2) * \pi r^2 = (1/2) * (22/7) * (7)^2 = (1 * 22 * 49) / (2 * 7) = 77$ m²

Area of Type B = $96 + 112 + 77 = 285$ m²

Converting this area to ft²: 285 m² = 285•10.76 ft² = 3066.6 ft²

Type B is (3066.6 - 1300 = 1766.6 ft²) 1766.6 ft² larger than type A.

24. A

The formula of the volume of cylinder is the base area multiplied by the height. As the formula:

Volume of a cylinder = $\pi r^2 h$. Where π is 3.142, r is radius of the cross sectional area, and h is the height.

We know that the diameter is 5 meters, so the

radius is 5/2 = 2.5 meters.

The volume is: V = 3.142 * 2.5² * 12 = 235.65 m³.

25. C
The large cube is made up of 8 smaller cubes with 5 cm sides. The volume of a cube is found by the third power of the length of one side.
Volume of the large cube = Volume of the small cube•8

= (5³) * 8 = 125 * 8

= 1000 cm³

There is another solution for this question. Find the side length of the large cube. There are two cubes rows with 5 cm length for each. So, one side of the large cube is 10 cm.

The volume of this large cube is equal to 10³ = 1000 cm³

26. A
The line is pointing towards numbers greater than 2. The equation is therefore, X > 2.

27. C
Pythagorean Theorem:
(Hypotenuse)² = (Perpendicular)² + (Base)²

h² = a² + b²

Given: a = 6, h = 10

h² = a² + b²

b² = h² - a²

b² = 10² + 6²

$b^2 = 100 - 36$

$b^2 = 64$
$b = 8$

28. D
Two parallel lines intersected by a third line with angles of 75°
$x = 75°$ (corresponding angles)
$x + y = 180°$ (supplementary angles)

$y = 180° - 75°$

$y = 105°$

29. D
The distance between two points is found by $= [(x_2 - x_1)^2 + (y_2 - y_1)^2]^{1/2}$

In this question:

$(18, 12) : x_1 = 18, y_1 = 12$

$(9, -6) : x_2 = 9, y_2 = -6$

Distance $= [(9 - 18)^2 + (-6 - 12)^2]^{1/2}$

$= [(-9)^2 + (-18)^2]^{1/2}$

$= (9^2 + 2^2 \cdot 9^2)^{1/2}$

$= (9^2(1 + 5))^{1/2}$ ... We can take 9 out of the square root:

$= 9 * 6^{1/2}$

$= 9\sqrt{6}$

$= 9 * 2.45$

$= 22.04$

The distance is approximately 22 units.

BC Police Math Practice

30. D
We have a circle given with diameter 8 cm and a square located within the circle. We are asked to find the area of the circle for which we only need to know the length of the radius that is the half of the diameter.
Area of circle = πr^2 ... r = 8/2 = 4 cm

Area of circle = $\pi * 4^2$

= 16π cm² ... As we notice, the inner square has no role in this question.

31. B
Perimeter of a parallelogram is the sum of the sides.
Perimeter = 2(l + b)
Perimeter = 2(3 +10), 2 x 13
Perimeter = 26 cm.

32. C
Volume of a cylinder is π x r^2 x h

Diameter = 5 ft. so radius is 2.5 ft.

Volume of cylinder= π x 2.5^2 x 2

= π x 6.25 x 2 = 12.5 π

Approximate π to 3.142

Volume of the cylinder = 39.25

Volume of a rectangle = height X width X length.
= 5 X 5 X 4 = 100

Total volume = Volume of rectangular solid + volume of cylinder

Total volume = 100 + 39.25

Total volume = 139.25 ft³ or about 140 ft³

33. B

If we know the coordinates of two points on a line, we can find the slope (m) with the below formula: $m = (y_2 - y_1)/(x_2 - x_1)$ where (x_1, y_1) represent the coordinates of one point and (x_2, y_2) the other.

In this question:

$(-4, -4) : x_1 = -4, y_1 = -4$

$(-1, 2) : x_2 = -1, y_2 = 2$

Inserting these values into the formula:

$m = (2 - (-4))/(-1 - (-4)) = (2 + 4)/(-1 + 4) = 6/3$ ... Simplifying by 3:

$m = 2$

34. C

Pythagorean Theorem:
(Hypotenuse)² = (Perpendicular)² + (Base)²

$h^2 = a^2 + b^2$

Given: $3^2 + 4^2 = h^2$

$h^2 = 9 + 16$

$h = \sqrt{25}$

$h = 5$

Basic Algebra

Basic algebra questions are included in the Patterns and Algebra Section of the math test.

Algebra questions cover:

- Ratio and proportion
- Linear equations
- Patterns
- Solve quadratics
- Solve real-world quadratic questions
- Identify quadratic equations from graphs
- Identify linear equations from graphs

Solving One-Variable Linear Equations

Linear equations with variable x is an equation with the following form:
$$ax = b$$

where a and b are real numbers. If a=0 and b is different from 0, then the equation has no solution.

Let's solve one simple example of a linear equation with one variable:

Algebra

$$4x - 2 = 2x + 6$$

When given this type of equation, move variables to one side, and real numbers to the other. Always remember: if you are changing sides, you are changing signs. Move all variables to the left, and real numbers to the right:
4x - 2 = 2x + 6
4x - 2x = 6 + 2
2x = 8
x = 8/2
x = 4

When 2x goes to the left it becomes -2x, and -2 goes to the right and becomes +2. After calculations, we find that x is 4, which is a solution of our linear equation.

Let's solve a little more complex linear equation:

2x - 6/4 + 4 = x
2x - 6 + 16 = 4x
2x - 4x = -16 + 6
-2x = -10
x = -10/-2
x = 5

We multiply whole equation by 4, to lose the fractional line. Now we have a simple linear equation. If we change sides, we change the signs.

Solving Two-Variable Linear Equations

If we have 2 or more linear equations with 2 or more variables, then we have a system of linear equations. The idea here is to express one variable using the other in one equation, and then use it in the second equation, so we get a linear equation with one variable. Here is an example:

x - y = 3
2x + y = 9

From the first equation, we express y using x.

y = x - 3

In the second equation, we write x-3 instead of y. And there we get a linear equation with one variable x.

2x + x - 3 = 9
3x = 9 + 3
3x = 12
x = 12/3
x = 4

Now that we found x, we can use it to find y.

y = x - 3
y = 4 - 3
y = 1

So, the solution of this system is (x,y) = (4,1).

Let's solve one more system using a different method:

Solve:

5x - 3y = 17
x + 3y = 11

5x - 3y + x + 3y = 17 - 11

Notice that we have -3y in the first equation and +3y in the second. If we add these 2, we get zero, which means we lose variable y. So, we add these 2 equations and we get a linear equation with one variable.

6x = 6
x = 1

Now that we have x, we use it to find y.

5 - 3y = 17
-3y = 17 - 5
-3y = 12
y = 12/(-3)
y = -4

Simplifying Polynomials

Let's say we are given some expression with one or more variables, where we have to add, subtract and multiply polynomials. We do the calculations with variables and constants and then we group the variables with the appropriate degrees. As a result, we would get a polynomial. This process is called simplifying polynomials, where we go from a complex expression to a simple polynomial.

Example:

Simplify the following expression and arrange the degrees from bigger to smaller:

$4 + 3x - 2x^2 + 5x + 6x^3 - 2x^2 + 1 = 6x^3 - 4x^2 + 8x + 5$

We can have more complex expressions such as:

$(x + 5)(1 - x) - (2x - 2) = x - x^2 + 5 - 5x - 2x + 2 = -x^2 - 6x + 7$

Here, first we multiply the polynomials and then we subtract the result and the third polynomial.

Factoring Polynomials

If we have a polynomial that we want to write as multiplication of a real number and a polynomial or as a multiplication of 2 or more polynomials, then we are dealing with factoring polynomials.

Let's see an example for a simple factoring:

$12x^2 + 6x - 4 =$
$2 * 6x^2 + 2 * 3x - 2 * 2 =$
$2(6x^2 + 3x - 2)$

We look at every polynomial member as a product of a real number and a variable. Notice that all real numbers in the polynomial are even, so they have the same number (factor). We pull out that 2 in front of the polynomial, and we write what is left.

What if have a more complex case, where we can't find a factor that is a real number? Here is an example:

$x^2 - 2x + 1 =$
$x^2 - x - x + 1 =$
$x(x - 1) - (x - 1) =$
$(x - 1)(x - 1)$

We can write -2x as –x-x . Now we group first 2 members and we see that they have the same factor x, which we can pull in front of them. For the other

Algebra

2 members, we pull the minus in front of them, so we can get the same binomial that we got with the first 2 members. Now we have that this binomial is the factor for x(x-1) and (x-1).

If we pull x-1 in front (underlined), from the first member we are left with x, and from the second we have -1.
And that is how we transform a polynomial into a product of 2 polynomials (in this case binomials).

Quadratic Equations

A. Factoring

Quadratic equations are usually called second degree equations, which mean that the second degree is the highest degree of the variable that can be found in the quadratic equation. The form of these equations is:

$ax^2 + bx + c = 0$

where a, b and c are some real numbers.

One way for solving quadratic equations is the factoring method, where we transform the quadratic equation into a product of 2 or more polynomials. Let's see how that works in one simple example:

x2 + 2x = 0
x(x + 2) = 0
(x = 0) V (x + 2 = 0)
(x = 0 V (x + -2)

Notice that here we don't have parameter c, but this is still a quadratic equation, because we have the

second degree of variable x. Our factor here is x, which we put in front, and we are left with x+2. The equation is equal to 0, so either x or x+2 are 0, or both are 0.
So, our 2 solutions are 0 and -2.

B. Quadratic formula
If we are unsure how to rewrite quadratic equations so we can solve it using factoring method, we can use the formula for quadratic equation:

$$x_{1,2} = \frac{-b \pm \sqrt{b^2 - 4ac}}{2a}$$

We write $x_{1,2}$ because it represents 2 solutions of the equation. Here is one example:

$3x^2 - 10x + 3 = 0$

$x_{1,2} = \frac{-b \pm \sqrt{b^2 - 4ac}}{2a}$

$x_{1,2} = \frac{-(-10) \pm \sqrt{(-10)^2 - 4 \cdot 3 \cdot 3}}{2 \cdot 3}$

$x_{1,2} = \frac{10 \pm \sqrt{100 - 36}}{6}$

$x_{1,2} = \frac{10 \pm \sqrt{64}}{6}$

$x_{1,2} = \frac{10 \pm 8}{6}$

$x_1 = \frac{10 + 8}{6} = \frac{18}{6} = 3$

$x_2 = \frac{10 - 8}{6} = \frac{2}{6} = \frac{1}{3}$

We see that a is 3, b is -10 and c is 3.
We use these numbers in the equation and do some calculations.

Notice that we have + and -, so x_1 is for + and x_2 is for -, and that's how we get 2 solutions.

— Algebra —

Answer Sheet

1. (A) (B) (C) (D)   18. (A) (B) (C) (D)
2. (A) (B) (C) (D)   19. (A) (B) (C) (D)
3. (A) (B) (C) (D)   20. (A) (B) (C) (D)
4. (A) (B) (C) (D)   21. (A) (B) (C) (D)
5. (A) (B) (C) (D)   22. (A) (B) (C) (D)
6. (A) (B) (C) (D)   23. (A) (B) (C) (D)
7. (A) (B) (C) (D)   24. (A) (B) (C) (D)
8. (A) (B) (C) (D)   25. (A) (B) (C) (D)
9. (A) (B) (C) (D)   26. (A) (B) (C) (D)
10. (A) (B) (C) (D)   27. (A) (B) (C) (D)
11. (A) (B) (C) (D)   28. (A) (B) (C) (D)
12. (A) (B) (C) (D)   29. (A) (B) (C) (D)
13. (A) (B) (C) (D)   30. (A) (B) (C) (D)
14. (A) (B) (C) (D)   31. (A) (B) (C) (D)
15. (A) (B) (C) (D)   32. (A) (B) (C) (D)
16. (A) (B) (C) (D)   33. (A) (B) (C) (D)
17. (A) (B) (C) (D)   34. (A) (B) (C) (D)

BC Police Math Practice

1. Solve the linear equation: -x - 7 = -3x - 9

    a. -1
    b. 0
    c. 1
    d. 2

2. Solve the system: 4x - y = 5   x + 2y = 8

    a. (3,2)
    b. (3,3)
    c. (2,3)
    d.   (2,2)

3. Simplify the following expression:

$3x^3 + 2x^2 + 5x - 7 + 4x^2 - 5x + 2 - 3x^3$

    a. $6x^2 - 9$
    b. $6x^2 - 5$
    c. $6x^2 - 10x - 5$
    d. $6x^2 + 10x - 9$

4. Find 2 numbers that sum to 21 and the sum of the squares is 261.

    a. 14 and 7
    b. 15 and 6
    c. 16 and 5
    d. 17 and 4

Algebra

5. Using the factoring method, solve the quadratic equation: $x^2 + 4x + 4 = 0$

    a. 0 and 1
    b. 1 and 2
    c. 2
    d. -2

6. Using the quadratic formula, solve the quadratic equation: $x - 31/x = 0$

    a. $-\sqrt{13}$ and $\sqrt{13}$
    b. $-\sqrt{31}$ and $\sqrt{31}$
    c. $-\sqrt{31}$ and $2\sqrt{31}$
    d. $-\sqrt{3}$ and $\sqrt{3}$

7. Using the factoring method, solve the quadratic equation: $2x^2 - 3x = 0$

    a. 0 and 1.5
    b. 1.5 and 2
    c. 2 and 2.5
    d. 0 and 2

8. Using the quadratic formula, solve the quadratic equation: $x^2 - 9x + 14 = 0$

    a. 2 and 7
    b. -2 and 7
    c. -7 and -2
    d. -7 and 2

9. Solve the following equation $4(y + 6) = 3y + 30$

    a. $y = 20$
    b. $y = 6$
    c. $y = 30/7$
    d. $y = 30$

10. Using the factoring method, solve the quadratic equation: $x^2 - 5x - 6 = 0$

    a. -6 and 1
    b. -1 and 6
    c. 1 and 6
    d. -6 and -1

11. Factor the polynomial $x^3y^3 - x^2y^8$.

    a. $x^2y^3(x - y^5)$
    b. $x^3y^3(1 - y^5)$
    c. $x^2y^2(x - y^6)$
    d. $xy^3(x - y^5)$

12. Find the solution for the following linear equation: $5x/2 = (3x + 24)/6$

    a. -1
    b. 0
    c. 1
    d. 2

Algebra

13. Solve the system, if a is some real number:

$ax + y = 1$
$x + ay = 1$

    a. $(1, a)$
    b. $(1/a + 1, 1)$
    c. $(1/(a + 1), 1/(a + 1))$
    d. $(a, 1/a + 1)$

14. Solve $3(x + 2) - 2(1 - x) = 4x + 5$

    a. -1
    b. 0
    c. 1
    d. 2

15. Simplify $3x^a + 6a^x - x^a + (-5a^x) - 2x^a$

    a. $a^x + x^a$
    b. $a^x - x^a$
    c. a^x
    d. x^a

16. A map uses a scale of 1:100,000. How much distance on the ground is 3 inches on the map if the scale is in inches?

    a. 13 inches
    b. 300,000 inches
    c. 30,000 inches
    d. 333.999 inches

BC Police Math Practice

17. Using the quadratic formula, solve the quadratic equation: $0.9x^2 + 1.8x - 2.7 = 0$

    a. 1 and 3
    b. -3 and 1
    c. -3 and -1
    d. -1 and 3

18. Find x and y from the following system of equations:

$(4x + 5y)/3 = ((x - 3y)/2) + 4$
$(3x + y)/2 = ((2x + 7y)/3) - 1$

    a. (1, 3)
    b. (2, 1)
    c. (1, 1)
    d. (0, 1)

19. Using the factoring method, solve the quadratic equation: $x^2 + 12x - 13 = 0$

    a. -13 and 1
    b. -13 and -1
    c. 1 and 13
    d. -1 and 13

Algebra

20. Using the quadratic formula, solve the quadratic equation: $((x^2 + 4x + 4) + (x^2 - 4x + 4)) / (x^2 - 4) = 0$.

    a. It has infinite numbers of solutions

    b. 0 and 1

    c. It has no solutions

    d. 0

21. Turn the following expression into a simple polynomial: $5(3x^2 - 2) - x^2(2 - 3x)$

    a. $3x^3 + 17x^2 - 10$

    b. $3x^3 + 13x^2 + 10$

    c. $-3x^3 - 13x^2 - 10$

    d. $3x^3 + 13x^2 - 10$

22. Solve $(x^3 + 2)(x^2 - x) - x^5$.

    a. $2x^5 - x^4 + 2x^2 - 2x$

    b. $-x^4 + 2x^2 - 2x$

    c. $-x^4 - 2x^2 - 2x$

    d. $-x^4 + 2x^2 + 2x$

23. $9ab^2 + 8ab^2 =$

    a. ab^2

    b. $17ab^2$

    c. 17

    d. $17a^2b^2$

BC Police Math Practice

24. Factor the polynomial $x^2 - 7x - 30$.

    a. $(x + 15)(x - 2)$
    b. $(x + 10)(x - 3)$
    c. $(x - 10)(x + 3)$
    d. $(x - 15)(x + 2)$

25. If a and b are real numbers, solve the following equation: $(a + 2)x - b = -2 + (a + b)x$

    a. -1
    b. 0
    c. 1
    d. 2

26. Turn the following expression into a simple polynomial: $1 - x(1 - x(1 - x))$

    a. $x^3 + x^2 - x + 1$
    b. $-x^3 - x^2 + x + 1$
    c. $-x^3 + x^2 - x + 1$
    d. $x^3 + x^2 - x - 1$

27. $7(2y + 8) + 1 - 4(y + 5) =$

    a. $10y + 36$
    b. $10y + 77$
    c. $18y + 37$
    d. $10y + 37$

Algebra

28. Richard gives 's' amount of salary to each of his 'n' employees weekly. If he has 'x' amount of money then how many days he can employ these 'n' employees.

    a. sx/7n
    b. 7x/nx
    c. nx/7s
    d. 7x/ns

29. Factor the polynomial $x^2 - 3x - 4$.

    a. $(x + 1)(x - 4)$
    b. $(x - 1)(x + 4)$
    c. $(x - 1)(x - 4)$
    d. $(x + 1)(x + 4)$

30. Using the quadratic formula, solve the quadratic equation:

$(a^2 - b^2)x^2 + 2ax + 1 = 0$

    a. $a/(a + b)$ and $b/(a + b)$
    b. $1/(a + b)$ and $a/(a + b)$
    c. $a/(a + b)$ and $a/(a - b)$
    d. $-1/(a + b)$ and $-1/(a - b)$

BC Police Math Practice

31. Turn the following expression into a simple polynomial:

$(a + b)(x + y) + (a - b)(x - y) - (ax + by)$

    a. $ax + by$
    b. $ax - by$
    c. $ax^2 + by^2$
    d. $ax^2 - by^2$

32. The area of a rectangle is 20 cm². If one side increases by 1 cm and other by 2 cm, the area of the new rectangle is 35 cm². Find the sides of the original rectangle.

    a. (4,8)
    b. (4,5)
    c. (2.5,8)
    d. b and c

33. Find the x-intercepts of the quadratic function
$f(x) = (x - 5)^2 - 9$.

    a. {2,4}
    b. {2,8}
    c. {4,8}
    d. {1,2}

Algebra

34. In a store, the price of t-shirts and pants are constant. If John buys 4 t-shirts and 5 pair of pants, he pays $51. If he buys 7 t-shirts and 3 pair of pants, then he pays $49. Find the difference between the price of one pair of pants and one t-shirt.

    a. 0
    b. 3
    c. 7
    d. 12

Answer Key

1. A

We should collect similar terms on the same side. Here, we can collect x terms on left side, and the constants on the right side:

- x - 7 = - 3x - 9 .... Let us add 3x to both sides:

- x - 7 + 3x = - 3x - 9 + 3x

2x - 7 = - 9 ... Now, we can add + 7 to both sides:

2x - 7 + 7 = - 9 + 7

2x = - 2 ... Dividing both sides by 2 gives us the value of x:

x = -2/2

x = -1

2. C

First, we need to write two equations separately:
4x - y = 5 (I)

x + 2y = 8 (II) ... Here, we can use two ways to solve the system. One is substitution method, the other one is linear elimination method:

1. Substitution Method

Equation (I) gives us that y = 4x - 5. We insert this value of y into equation (II):

x + 2(4x - 5) = 8

x + 8x - 10 = 8

9x - 10 = 8

Algebra

$9x = 18$

$x = 2$

Bu knowing x = 2, we can find the value of y by inserting x = 2 into either of the equations. Let us choose equation (I):

$4(2) - y = 5$

$8 - y = 5$

$8 - 5 = y$

$y = 3 \rightarrow$ solution is (2, 3)

2. Linear Elimination Method:

2•/ 4x - y = 5 ... by multiplying equation (I) by 2, we see that -2y will form; and y terms

  x + 2y = 8 ... will be eliminated when summed with +2y in equation (II):

2•/ 4x - y = 5

+   x + 2y = 8

   8x - 2y = 10

 + x + 2y = 8 ... Summing side by side:

$8x + x - 2y + 2y = 10 + 8$ ... -2y and +2y eliminate each other:

$9x = 18$

$x = 2$

By knowing x = 2, we can find the value of y by inserting x = 2 into either of the equations. Let us choose equation (I):

$4(2) - y = 5$

$8 - y = 5$

$8 - 5 = y$

$y = 3 \rightarrow$ solution is (2, 3)

3. B

$3x^3 + 2x^2 + 5x - 7 + 4x^2 - 5x + 2 - 3x^3$ ... write similar terms together:

$= 3x^3 - 3x^3 + 2x^2 + 4x^2 + 5x - 5x - 7 + 2$ ... operate within the same terms. $3x^3$ and $-3x^3$, $5x$ and $-5x$ cancel:

$= 6x^2 - 5$

4. B

There are two statements made. This means that we can write two equations according to these statements:

The sum of two numbers are 21: $x + y = 21$

The sum of the squares is 261: $x^2 + y^2 = 261$

We are asked to find x and y.

Since we have the sums of the numbers and the sums of their squares; we can use the square formula of $x + y$, that is:

$(x + y)^2 = x^2 + 2xy + y^2$ ... Here, we can insert the known values $x + y$ and $x^2 + y^2$:

$(21)^2 = 261 + 2xy$ ... Arranging to find xy:

$441 = 261 + 2xy$

$441 - 261 = 2xy$

180 = 2xy

xy = 180/2

xy = 90

We need to find two numbers which multiply to 90. Checking the answer choices, we see that in (b), 15 and 6 are given. 15•6 = 90. Also their squares sum up to 261 ($15^2 + 6^2 = 225 + 36 = 261$). So these two numbers satisfy the equation.

5. D
$x^2 + 4x + 4 = 0$ ... We try to separate the middle term 4x to find common factors with x^2 and 4 separately:

$x^2 + 2x + 2x + 4 = 0$ ... Here, we see that x is a common factor for x^2 and 2x, and 2 is a common factor for 2x and 4:

$x(x + 2) + 2(x + 2) = 0$ ... Here, we have x times x + 2 and 2 times x + 2 summed up. This means that we have x + 2 times x + 2:

$(x + 2)(x + 2) = 0$

$(x + 2)^2 = 0$ ... This is true if only if x + 2 is equal to zero.

x + 2 = 0

x = -2

6. B
To solve the equation, first we need to arrange it to appear in the form $ax^2 + bx + c = 0$ by removing the denominator:

x - 31/x = 0 ... First, we enlarge the equation by x:

x•x - 31•x/x = 0

x² - 31 = 0

The quadratic formula to find the roots of a quadratic equation is:

$x_{1,2}$ = (-b ± √Δ) / 2a where Δ = b² - 4ac and is called the discriminant of the quadratic equation.

In our question, the equation is x² - 31 = 0. By remembering the form ax² + bx + c = 0:

a = 1, b = 0, c = -31

So, we can find the discriminant first, and then the roots of the equation:

Δ = b² - 4ac = 0² - 4•1•(-31) = 124

$x_{1,2}$ = (-b ± √Δ) / 2a = (±√124) / 2 = (±√4•31) / 2 = (±2√31) / 2 ... Simplifying by 2:

$x_{1,2}$ = ±√31 ... This means that the roots are √31 and -√31.

7. A
2x² - 3x = 0 ... we see that both of the terms contain x; so we can take it out as a factor:

x(2x - 3) = 0 ... two terms are multiplied and the result is zero. This means that either of the terms or, both can be equal to zero:

x = 0 ... this is one solution

2x - 3 = 0 → 2x = 3 → x = 3/2 → x = 1.5 ... this is the second solution.

So, the solutions are 0 and 1.5.

8. A

To solve the equation, we need the equation in the form $ax^2 + bx + c = 0$.

$x^2 - 9x + 14 = 0$ is already in this form.

The quadratic formula to find the roots of a quadratic equation is:

$x_{1,2} = (-b \pm \sqrt{\Delta}) / 2a$ where $\Delta = b^2 - 4ac$ and is called the discriminant of the quadratic equation.

In our question, the equation is $x^2 - 9x + 14 = 0$. By remembering the form $ax^2 + bx + c = 0$:

$a = 1, b = -9, c = 14$

So, we can find the discriminant first, and then the roots of the equation:

$\Delta = b^2 - 4ac = (-9)^2 - 4 \cdot 1 \cdot 14 = 81 - 56 = 25$

$x_{1,2} = (-b \pm \sqrt{\Delta}) / 2a = (-(-9) \pm \sqrt{25}) / 2 = (9 \pm 5) / 2$

This means that the roots are,

$x_1 = (9 - 5) / 2 = 2$ and $x_2 = (9 + 5) / 2 = 7$

9. B

$4y + 24 = 3y + 30$, $= 4y - 3y + 24 = 30$, $= y + 24 = 30$, $= y = 30 - 24$, $= y = 6$

10. B

$x^2 - 5x - 6 = 0$

We try to separate the middle term $-5x$ to find common factors with x^2 and -6 separately:

$x^2 - 6x + x - 6 = 0$ ... Here, we see that x is a common factor for x^2 and $-6x$:

$x(x - 6) + x - 6 = 0$ ... Here, we have x times x - 6 and 1 time x - 6 summed up. This means that we have x + 1 times x - 6:

$(x + 1)(x - 6) = 0$ ... This is true when either or both of the expressions in the parenthesis are equal to zero:

$x + 1 = 0$ ... $x = -1$

$x - 6 = 0$ ... $x = 6$

-1 and 6 are the solutions for this quadratic equation.

11. A
We need to find the greatest common divisor of the two terms to factor the expression. We should remember that if the bases of exponent numbers are the same, the multiplication of two terms is found by summing the powers and writing on the same base. Similarly; when dividing, the power of the divisor is subtracted from the power of the divided.

Both x^3y^3 and x^2y^8 contain x^2 and y^3. So;

$x^3y^3 - x^2y^8 = x \cdot x^2y^3 - y^5 \cdot x^2y^3$ ... We can carry x^2y^3 out as the factor:

$= x^2y^3(x - y^5)$

12. D
Our aim to collect the knowns on one side and the unknowns (x terms) on the other side:

$5x/2 = (3x + 24)/6$ ... First, we can simplify the denominators of both sides by 2:

$5x = (3x + 24)/3$ ... Now, we can cross multiply:

$15x = 3x + 24$

$15x - 3x = 24$

$12x = 24$

$x = 24/12 = 2$

13. C
Solving the system means finding x and y. Since we also have a in the system, we will find x and y depending on a.

We can obtain y by using the equation $ax + y = 1$:

$y = 1 - ax$ ... Then, we can insert this value into the second equation:

$x + a(1 - ax) = 1$

$x + a - a^2x = 1$

$x - a^2x = 1 - a$

$x(1 - a^2) = 1 - a$ ... We need to obtain x alone:

$x = (1 - a)/(1 - a^2)$ ... Here, $1 - a^2 = (1 - a)(1 + a)$ is used:

$x = (1 - a)/((1 - a)(1 + a))$ ... Simplifying by $(1 - a)$:

$x = 1/(a + 1)$ ... Now we know the value of x. By using either of the equations, we can find the value of y. Let us use $y = 1 - ax$:

$y = 1 - a \cdot 1/(a + 1)$

$y = 1 - a/(a + 1)$ ... By writing on the same denominator:

$y = ((a + 1) - a)/(a + 1)$

$y = (a + 1 - a)/(a + 1)$ ... a and -a cancel each other:

$y = 1/(a + 1)$ ... x and y are found to be equal.

The solution of the system is $(1/(a + 1), 1/(a + 1))$

14. C
To solve the linear equation, we operate the knowns and unknowns within each other and try to obtain x term (which is the unknown) alone on one side of the equation:
$3(x + 2) - 2(1 - x) = 4x + 5$ ... We remove the parenthesis by distributing the factors:

$3x + 6 - 2 + 2x = 4x + 5$

$5x + 4 = 4x + 5$

$5x - 4x = 5 - 4$

$x = 1$

15. C
Here, we use the commutative property of multiplication, meaning that $xa = ax$:
$3xa + 6ax - xa + (-5ax) - 2xa = 3ax + 6ax - ax - 5ax - 2ax$
$= (3 + 6 - 1 - 5 - 2)ax$
$= (9 - 8)ax$
$= ax$

16. B
1 inch on map = 100,000 inches on ground. So 3 inches on map = 3 x 100,000 = 300,000 inches on ground.

17. B

To solve the equation, we need the equation in the form $ax^2 + bx + c = 0$.

$0.9x^2 + 1.8x - 2.7 = 0$ is already in this form.

The quadratic formula to find the roots of a quadratic equation is:

$x_{1,2} = (-b \pm \sqrt{\Delta}) / 2a$ where $\Delta = b^2 - 4ac$ and is called the discriminant of the quadratic equation.

In our question, the equation is $0.9x^2 + 1.8x - 2.7 = 0$. To eliminate the decimals, let us multiply the equation by 10:
$9x^2 + 18x - 27 = 0$ ... This equation can be simplified by 9
since each term contains 9:

$x^2 + 2x - 3 = 0$

By remembering the form $ax^2 + bx + c = 0$:

$a = 1, b = 2, c = -3$

So, we can find the discriminant first, and then the roots of the equation:

$\Delta = b^2 - 4ac = (2)^2 - 4 \bullet 1 \bullet (-3) = 4 + 12 = 16$

$x_{1,2} = (-b \pm \sqrt{\Delta}) / 2a = (-2 \pm \sqrt{16}) / 2 = (-2 \pm 4) / 2$

This means that the roots are,

$x_1 = (-2 - 4)/2 = -3$ and $x_2 = (-2 + 4)/2 = 1$

18. C
First, we need to arrange the two equations to ob-

tain the form ax + by = c. We see that there are 3 and 2 in the denominators of both equations. If we equate all at 6, then we can cancel all 6 in the denominators and have straight equations:

Equate all denominators at 6:

2(4x + 5y)/6 = 3(x - 3y)/6 + 4•6/6 ... Now we can cancel 6 in the denominators:

8x + 10y = 3x - 9y + 24 ... We can collect x and y terms on left side of the equation:

8x + 10y - 3x + 9y = 24

5x + 19y = 24 ... Equation (I)
Arrange the second equation:

3(3x + y)/6 = 2(2x + 7y)/6 - 1•6/6 ... Now we can cancel 6 in the denominators:

9x + 3y = 4x + 14y - 6 ... We can collect x and y terms on left side of the equation:

9x + 3y - 4x - 14y = -6

5x - 11y = -6 ... Equation (II)

Now, we have two equations and two unknowns x and y. By writing the two equations one under the other and operating, we can find one unknown first, and find the other next:

    5x + 19y = 24
-1/ 5x - 11y = -6 ... If we substitute this equation from the upper one, 5x cancels -5x:

$5x + 19y = 24$

$\underline{-5x + 11y = 6}$ ... Summing side-by-side:

$5x - 5x + 19y + 11y = 24 + 6$

$30y = 30$ ... Dividing both sides by 30:
$y = 1$

Inserting $y = 1$ into either of the equations, we can find the value of x. Choosing equation I:

$5x + 19 \cdot 1 = 24$

$5x = 24 - 19$
$5x = 5$ ... Dividing both sides by 5:

$x = 1$

So, $x = 1$ and $y = 1$ is the solution; shown as (1, 1).

19. A
$x^2 + 12x - 13 = 0$ ... We try to separate the middle term 12x to find common factors with x^2 and -13 separately:

$x^2 + 13x - x - 13 = 0$ ... Here, we see that x is a common factor for x^2 and 13x, and -1 is a common factor for -x and -13:

$x(x + 13) - 1(x + 13) = 0$ ... Here, we have x times x + 13 and -1 times x + 13 summed up. This means that we have x - 1 times x + 13:

$(x - 1)(x + 13) = 0$

This is true when either, or both, the expressions in the parenthesis are equal to zero:

x - 1 = 0 ... x = 1

x + 13 = 0 ... x = -13

1 and -13 are the solutions for this quadratic equation.

20. C
First, we need to simplify the equation:
$((x^2 + 4x + 4) + (x^2 - 4x + 4)) / (x^2 - 4) = 0$

$(x^2 + 4x + 4 + x^2 - 4x + 4) / (x^2 - 4) = 0$ ... 4x and -4x in the numerator cancel.

Note that $x^2 - 4$ is two square difference and is equal to $x^2 - 2^2 = (x - 2)(x + 2)$:

$(2x^2 + 8)/((x - 2)(x + 2)) = 0$

The denominator tells us that if x - 2 or x + 2 equals to zero, there will be no solution. So, we will need to eliminate x = 2 and x = -2 from our solution which will be found considering the numerator:

$2x^2 + 8 = 0$

$2(x^2 + 4) = 0$

$x^2 + 4 = 0$

$x^2 = -4$ ... We know that, a square cannot be equal to a negative number. Solution for the square root of -4 is not a real number, so this equation has no solution.

21. D
We need to distribute the factors to the terms inside the related parenthesis:

$5(3x^2 - 2) - x^2(2 - 3x) = 15x^2 - 10 - (2x^2 - 3x^3)$

$= 15x^2 - 10 - 2x^2 + 3x^3$

$= 3x^3 + 15x^2 - 2x^2 - 10$ ... similar terms written together to ease summing/substituting.

$= 3x^3 + 13x^2 - 10$

22. B
We need to distribute the factors to the terms inside the related parenthesis:

$(x^3 + 2)(x^2 - x) - x^5 = x^5 - x^4 + (2x^2 - 2x) - x^5$

$= x^5 - x^4 + 2x^2 - 2x - x^5$

$= x^5 - x^5 - x^4 + 2x^2 - 2x$ ... similar terms written together to ease summing/substituting.

$= -x^4 + 2x^2 - 2x$

23. B
To simplify the expression, we need to find common factors. We see that both terms contain the term ab^2. So, we can take this term out of each term as a factor:

$9ab^2 + 8ab^2 = (9 + 8) ab^2 = 17ab^2$

24. C
$x^2 - 7x - 30 = 0$ ... We try to separate the middle term $-7x$ to find common factors with x^2 and -30 separately:

$x^2 - 10x + 3x - 30 = 0$ ... Here, we see that x is a common factor for x^2 and $-10x$, and 3 is a common factor for $3x$ and -30:

$x(x - 10) + 3(x - 10) = 0$ ... Here, we have x times x - 10 and 3 times x - 10 summed up. This means that we have x + 3 times x - 10:

$(x + 3)(x - 10) = 0$ or $(x - 10)(x + 3) = 0$

25. A

We need to simplify the equation by distributing factors and then collecting x terms on one side, and the others on the other side:

$(a + 2)x - b = -2 + (a + b)x$

$ax + 2x - b = -2 + ax + bx$

$ax + 2x - ax - bx = -2 + b$ ... ax and -ax cancel each other:

$2x - bx = -2 + b$ ... we take -1 as a factor on the right side:

$(2 - b)x = -(2 - b)$

$x = -(2 - b)/(2 - b)$ ... Simplifying by 2 - b:

$x = -1$

26. C

To obtain a polynomial, remove the parenthesis by distributing the related factors to the terms inside the parenthesis:
$1 - x(1 - x(1 - x)) = 1 - x(1 - (x - x \cdot x)) = 1 - x(1 - x + x^2)$

$= 1 - (x - x \cdot x + x \cdot x^2) = 1 - x + x^2 - x^3$ ... Writing this result in descending order of powers:

$= - x^3 + x^2 - x + 1$

27. D

To simplify the expression, remove the parenthesis by distributing the related factors to the terms inside the parenthesis:

$7(2y + 8) + 1 - 4(y + 5) = (7 \cdot 2y + 7 \cdot 8) + 1 - (4 \cdot y + 4 \cdot 5)$

$= 14y + 56 + 1 - 4y - 20$

$= 14y - 4y + 56 + 1 - 20$ ... similar terms written together to ease summing/substituting.

$= 10y + 37$

28. D

We are given that each of the n employees earns s amount of salary weekly. This means that one employee earns s salary weekly. So; Richard has ns amount of money to employ n employees for a week.

We are asked to find the number of days n employees can be employed with x amount of money. We can do simple direct proportion:

If Richard can employ n employees for 7 days with ns amount of money,

Richard can employ n employees for y days with x amount of money ... y is the number of days we need to find.

Cross multiply:

$y = (x \cdot 7)/(ns)$

$y = 7x/ns$

29. A

$x^2 - 3x - 4$ ... try to separate the middle term $-3x$ to find common factors with x^2 and -4 separately:

$x^2 + x - 4x - 4$ ... Here, x is a common factor for x^2 and x, and -4 is a common factor for $-4x$ and -4:

$= x(x + 1) - 4(x + 1)$ ... Here, x times $x + 1$ and -4 times x + 1 summed up. This means that we have x - 4 times x + 1:

= (x - 4)(x + 1) or (x + 1)(x - 4)

30. D

To solve the equation, we need the equation in the form $ax^2 + bx + c = 0$.

$(a^2 - b^2)x^2 + 2ax + 1 = 0$ is already in this form.

The quadratic formula to find the roots of a quadratic equation is:

$x_{1,2} = (-b \pm \sqrt{\Delta}) / 2a$ where $\Delta = b^2 - 4ac$ and is called the discriminant of the quadratic equation.

In our question, the equation is $(a^2 - b^2)x^2 + 2ax + 1 = 0$.

By remembering the form $ax^2 + bx + c = 0$: $a = a^2 - b^2$, $b = 2a$, $c = 1$

So, we can find the discriminant first, and then the roots of the equation:

$\Delta = b^2 - 4ac = (2a)^2 - 4(a^2 - b^2) * 1 = 4a^2 - 4a^2 + 4b^2 = 4b^2$

$x_{1,2} = (-b \pm \sqrt{\Delta}) / 2a = (-2a \pm \sqrt{4b^2}) / (2(a^2 - b^2)) = (-2a \pm 2b) / (2(a^2 - b^2))$

$= 2(-a \pm b) / (2(a^2 - b^2))$ ... We can simplify by 2:

$= (-a \pm b) / (a^2 - b^2)$

This means that the roots are,

$x_1 = (-a - b) / (a^2 - b^2)$ ... $a^2 - b^2$ is two square differences:

$x_1 = -(a + b) / ((a - b)(a + b))$ ... $(a + b)$ terms cancel:

$x_1 = -1/(a - b)$

$x_2 = (-a + b) / (a^2 - b^2)$ ... $a^2 - b^2$ is two square differences:

$x_2 = -(a - b) / ((a - b)(a + b))$ ... $(a - b)$ terms cancel:

$x_2 = -1/(a + b)$

31. A

To simplify, remove the parenthesis and see if any terms cancel:

$(a + b)(x + y) + (a - b)(x - y) - (ax + by) = ax + ay + bx + by + ax - ay - bx + by - ax - by$

Writing similar terms together:

$= ax + ax - ax + bx - bx + ay - ay + by + by - by$ ... + terms cancel - terms:

$= ax + by$

32. D

The area of a rectangle is found by multiplying the width to the length. If we call these sides with "a" and "b"; the area is = a•b.

We are given that a•b = 20 cm² ... Equation I

One side is increased by 1 and the other by 2 cm. So new side lengths are "a + 1" and "b + 2."

The new area is $(a + 1)(b + 2) = 35$ cm² ... Equation II

Using equations I and II, we can find a and b:

ab = 20

BC Police Math Practice

$(a + 1)(b + 2) = 35$ ... distribute the terms in parenthesis:

$ab + 2a + b + 2 = 35$

Insert $ab = 20$ to the above equation:

$20 + 2a + b + 2 = 35$

$2a + b = 35 - 2 - 20$

$2a + b = 13$ ... This is one equation with two unknowns. We need to use another information to have two equations with two unknowns which leads us to the solution. We know that $ab = 20$. So, we can use $a = 20/b$:

$2(20/b) + b = 13$

$40/b + b = 13$ ... equate all denominators to "b" and eliminate it:

$40 + b^2 = 13b$

$b^2 - 13b + 40 = 0$ ... use the roots by factoring. We try to separate the middle term $-13b$ to find common factors with b^2 and 40 separately:

$b^2 - 8b - 5b + 40 = 0$ ... Here, b is a common factor for b^2 and $-8b$, and -5 is a common factor for $-5b$ and 40:

$b(b - 8) - 5(b - 8) = 0$ Here, b times $b - 8$ and -5 times $b - 8$ summed up. This means that we have $b - 5$ times $b - 8$:

$(b - 5)(b - 8) = 0$

This is true when either or both of the expressions in the parenthesis are equal to zero:

b - 5 = 0 ... b = 5

b - 8 = 0 ... b = 8

So we have two values for b which means we have two values for a as well. To find a, we can use any equation we have. Let us use a = 20/b.

If b = 5, a = 20/b → a = 4

If b = 8, a = 20/b → a = 2.5

So, (a, b) pairs for the sides of the original rectangle are: (4, 5) and (2.5, 8). These are found in (b) and (c) answer choices.

33. B
Finding the x-intercepts of a function means that we need to equate the function to zero and find the roots of the
equation:

$(x - 5)^2 - 9 = 9$

$(x - 5)^2 = 9$

$\sqrt{(x - 5)^2} = \sqrt{9}$

x - 5 = 3 → x = 8

x - 5 = -3 → x = 2

34. B
We have two variables: the price of a t-shirt and a pair of pants; and we have two situations given about them. We need to set two equations and solve them for the variables. Then, we are asked to find the difference.
Let us call the price of a t-shirt by a, and the price of a pair of pants by b:

BC Police Math Practice

If John buys 4 t-shirts and 5 pair of pants, he pays $51 → 4a + 5b = 51

If he buys 7 t-shirts and 3 pair of pants, then he pays
$49 → 7a + 3b = 49

4a + 5b = 51

7a + 3b = 49

We have two paths to follow: substitution or elimination. Here, since extracting a or b from either equation results in fractions; it is easier to choose elimination:

-3/ 4a + 5b = 51

 5/ 7a + 3b = 49

-12a - 15b = -153

 35a + 15b = 245
 ─────────────────
       23a = 92

        a = 4

Choosing either of the equations, find b, by inserting a:

4 * 4 + 5b = 51

16 + 5b = 51

5b = 35

b = 7

The difference between a and b is 7 - 4 = 3.

How to Answer Basic Math Questions

First, read the problem, but not the answers.

Work through the problem first and come up with your own answers. Hopefully, you should find your answer among the choices.

If no answer matches the one you got, re-check your math, but this time, use a different method. In math, there are different ways to solve a problem.

Math Multiple Choice Strategy

The two strategies for working with basic math multiple choice are Estimation and Elimination.

Estimation is just as it sounds - try to estimate an approximate answer first. Then look at the choices.

Elimination is probably the most powerful strategy for answering multiple choice.

Eliminate obviously incorrect answers and narrowing the possible choices.

Here are a few basic math examples of how this works.

BC Police Math Practice

Solve 2/3 + 5/12

- a. 9/17
- b. 3/11
- c. 7/12
- d. 1 1/12

First estimate the answer. 2/3 is more than half and 5/12 is about half, so the answer is going to be very close to 1.

Next, Eliminate. Choice A is about 1/2 and can be eliminated, choice B is very small, less than 1/2 and can be eliminated. Choice C is close to 1/2 and can be eliminated. Leaving only choice D, which is just over 1.

Work through the solution, find a common denominator and add. The correct answer is 1 1/12, so Choice D is correct.

Let's look at another example:

Solve 4/5 – 2/3

- a. 2/2
- b. 2/13
- c. 1
- d. 2/15

First, quickly estimate the answer. 4/5 is very close to 1, and 2/3 more than half, so the answer is going to be less than 1/2.

Basic Math Multiple Choice

Choice A can be eliminated right away, because it is 1. Choice C can be eliminated for the same reason.

Next, look at the denominators. Since 5 and 3 don't go into 13, choice B can be eliminated as well. That leaves choice D. Checking the answer, the common denominator will be 15. So the answer is 2/15 and choice D is correct.

Fractions Shortcut - Canceling Out

In any operation with fractions, if the numerator of one fraction has a common multiple with the denominator of the other, you can cancel out. This saves time and simplifies the problem quickly, making it easier to manage.

Solve 2/15 ÷ 4/5

    a.   6/65
    b.   6/75
    c. 5/12
    d.   1/6

To divide fractions, we multiply the first fraction with the inverse of the second fraction. Therefore we have 2/15 x 5/4. The numerator of the first fraction, 2, shares a multiple with the denominator of the second fraction, 4, which is 2. These cancel out, which gives, 1/3 x 1/2 = 1/6

Canceling out solved the questions very quickly, but we can still use multiple choice strategies to answer.

Choice B can be eliminated because 75 is too large a

denominator. Choice C can be eliminated because 5 and 15 don't go into 12.

Choice D is correct.

Basic Math Multiple Choice Strategy and Short-cuts

Multiplying decimals gives a very quick way to estimate and eliminate choices. Anytime that you multiply decimals, it is going to give an answer with the same number of decimal places as the combined operands.

So for example,

2.38 X 1.2 will produce a number with three places of decimal, which is 2.856.
Here are a few examples with step-by-step explanation:

Solve 2.06 x 1.2

    a.   24.82

    b.   2.482

    c. 24.72

    d.   2.472

This is a simple question, but even before you start calculating, you can eliminate several choices. When multiplying decimals, there will always be as many numbers behind the decimal place in the answer as the sum of the ones in the initial problem, so choices A and C can be eliminated.

How to Study for a Math Test

The correct answer is D: 2.06 x 1.2 = 2.472

Solve 20.0 ÷ 2.5

    a.   12.05
    b.   9.25
    c. 8.3
    d.   8

First estimate the answer to be around 10, and eliminate choice A. And since it'd also be an even number, you can eliminate choices B and C, leaving only choice D.

The correct Answer is D: 20.0 ÷ 2.5 = 8

How to Study for a Math Test

EVERY SUBJECT HAS ITS OWN PARTICULAR STUDY METHOD. Math is mostly numerical, not verbal and requires logical thinking; it has its own way to be studied. Before touching on significant points of studying a math test, lets look at some of the fundamentals of "learning."

Learning is not an instant experience; it is a procedure. Learning is a process not an event. Rome wasn't built in a day, and learning anything (or everything) isn't going to happen in a day either. You cannot expect to learn everything in one day, at night, before the test. It is important and necessary to learn day-by-day. Good time management plays a considerable role in learning. When you manage your time, and begin test preparation well in advance, you will notice the subjects are easier than you thought, or feared, and you will take the test without the stress of a sleepless body and an anxious mind.

Memorizing is a temporary step of learning if information is not comprehended and applied afterwards. Memorize just the basics and understand the meaning; then apply, analyze, synthesize and evaluate.

How to Study for a Math Test

These are the hierarchical layout of cognitive learning: Of course, there are some basic properties that you need to memorize in the beginning, since you cannot prove the facts every time you solve a math test. For example; the inner angles of a triangle sum up to 180°. If you do not know this, you may not be able to solve triangle problems. And, more importantly, if you do not practice, you will certainly forget this property. Practice helps information take root in your brain. Applying the same property on various types of questions extends the roots.

For example, if you see a triangle, you can analyze the question by means of the property. In a question, if you see a hexagon, you can split it into triangles and use the property, called synthesizing followed by evaluation. If all these steps are followed, the property is completely learned and has its place in your long term memory.

A useful method in providing consistent learning is using similarities between the information and events, images, shapes, ... etc. For example, assume that you have difficulty remembering the formula $x^a/y^b = x^a y^{-b} = 1/x^{-a} y^b$ in mind. You can associate this to an elevator: The exponents changing location (nominator/denominator) need to change exponent sign, similarly, people going up need to push the up button and if they decide to go down, they need to push the down button; so they need to change the button. Also; writing the formula in large letters and sticking it on a surface that is frequently visible helps memorizing it by using visual intelligence. The more senses (visual, musical, auditory, logical, ...) the material addresses to, the more permanent it is.

Attend to all classes. Knowledge is not replaceable by others, and every brain is unique. You cannot learn math from your classmate's notes; take your own notes in your own understanding of the material. What you understand, or don't understand, and how you understand it is different to everyone else. Highlight the important points in your own way. Remember math and all other courses are mostly learned at school -practice comes afterwards at home.

Find your own way of learning. Every person learns and studies differently. Some take notes, some do not like writing; listening is the major way of embracing information for them, and some watch. It is important to detect the way that is more useful for you. Coloring important points also helps. Due to selective perception; we see the attractive words, signs before the rest. While studying math, make a list; first, determine the subjects you feel inadequate on and focus on them initially.

Never gloss over something that you do not fully understand. Information is built on previous learning in a hierarchical order. If you have question marks about a mathematical property, and don't understand is completely, you cannot solve problems using that property. You need to have a strong background to succeed in math. You need to know your basic math inside out to do algebra. And you need to know your algebra inside out to do calculus. If you do not know exponentials, you cannot solve logarithms. If you don't understand something, get help from your teachers, reread course materials, resolve examples, discuss with friends or hire a private tutor. Never skip over something that you don't understand – it will come back to haunt you!

Practice makes perfect! Yes – it really does! Working through math problems in your own way is essential. Looking over examples is a good first step – but only that. The example solutions shown in the textbook show you have to solve a problem, next you have to do it yourself. Do not miss to do your assignments. You see different types of examples and acquire different outlooks when facing math problems. Math is fun because usually there are many ways to reach the solution. Find alternative ways to solve a problem which anchors the learning deeper. Find similar and different problems, discuss with friends, ask each other questions. Observing other people's way of thinking, and solving problems will help both of you improves your minds.

Succeeding in math is a mental action. However, do not disregard physical and psychological effects. Always think positive and never give up; no success is gained without effort. Of course you will waste time solving problems which you will find easy, and you will struggle with difficult problems. In the end, you will be one step further ahead. And after more time spent practicing, you will be another step ahead.

Reward yourself after intense studies. This will keep your motivation high. The reward may be a chocolate bar, playing a game for 20 minutes, or taking a walk in the park. It is very essential to have a good night's sleep before taking a math test. Eating habits directly effect success. Keep away from fast food as much as you can, eat a light meal before a test. At last but not least, keep your inner motivation very high. Believe in yourself; you will certainly get the good result of your planned, efficient studies.

How to Prepare for a Test

Most students hide their heads and procrastinate when faced with preparing for an examination, hoping that somehow they will be spared the agony of taking that test, especially if it is a big one that their futures rely on. Avoiding the all-important test is what many students do best and unfortunately, they suffer the consequences because of their lack of preparation.

Test preparation requires strategy. It also requires a dedication to getting the job done. It is the perfect training ground for anyone planning a professional life. Besides having several reliable strategies, the wise student also has a clear goal in mind and knows how to accomplish it. These tried and true concepts have worked well and will make your test preparation easier.

Test Prep and Study Skills Video Tutorials

https://www.test-preparation.ca/video-series-on-test-preparation-multiple-choice-strategies-and-how-to-study/

The Study Approach

Take responsibility for your own test preparation.

Prepare for a Test

It is a common- but big - mistake to link your studying to someone else's. Study partners are great, but only if they are reliable. It is your job to be prepared for the test, even if a study partner fails you. Do not allow others to distract you from your goals.

Prioritize the time available to study

When do you learn best, early in the day or in the dark of night? Does your mind absorb and retain information most efficiently in small blocks of time, or do you require long stretches to get the most done? It is important to figure out the best blocks of time available to you when you can be the most productive. Try to consolidate activities to allow for longer periods of study time.

Find a quiet place where you will not be disturbed

Do not try to squeeze in quality study time in any old location. Find a peaceful place with a minimum of distractions, such as the library, a park or even the laundry room. Good lighting is essential and you need to have comfortable seating and a desk surface large enough to hold your materials. It is probably not a great idea to study in your bedroom. You might be distracted by clothes on the floor, a book you have been planning to read, the telephone or something else. Besides, in the middle of studying, that bed will start to look very comfortable. Whatever you do, avoid using the bed as a place to study since you might fall asleep as a way of avoiding your work! That is the last thing that you should be doing during study time.

The exception is flashcards. By far the most productive study time is sitting down and studying and studying only. However, with flashcards you can carry them with you and make use of odd moments, like standing in line or waiting for the bus. This isn't as productive, but it really helps and is definitely worth doing.

Determine what you need to study

Gather together your books, your notes, your laptop and any other materials needed to focus on your study for this exam. Ensure you have everything you need so you don't waste time. Remember paper, pencils and erasers, sticky notes, bottled water and a snack. Keep your phone with you if you need it to find out essential information, but keep it turned off so others can't distract you.

Have a positive attitude

It is essential that you approach your studies for the test with an attitude that says you will pass it. And pass it with flying colors! This is one of the most important keys to successful study strategy. Believing that you are capable actually helps you to become capable.

The Strategy of Studying

Make materials easy to review and access

Consolidate materials to help keep your study area clutter free. If you have a laptop and a means of getting on line, you do not need a dictionary and thesaurus as well since those things are easily accessible via the internet. Go through written notes and consolidate those, as well. Have everything you need, but do not weigh yourself down with duplicates.

Review class notes

Stay on top of class notes and assignments by reviewing them frequently. Re-writing notes can be a terrific study trick, as it helps lock in information. Pay special attention to any comments that have been made by the teacher. If a study guide has been made available as part of the class materials, use it! It will be a valuable tool to use for studying.

Estimate how much time you will need

If you are concerned about the amount of time you have available it is a good idea to set up a schedule so that you do not get bogged down on one section and end without enough time left to study other things. Remember to schedule break time, and use that time for a little exercise or other stress reducing techniques.

Test yourself to determine your weaknesses

Look online for additional assessment and evaluation tools available for a particular subject. Visit our website https://www.test-preparation.ca for test tips and more practice questions. Once you have determined areas of concern, you will be able to focus on studying the information they contain and just brush up on the other areas of the exam.

Mental Prep – How to Psych Yourself Up for a Test

Because tests contribute mightily to your final class grade or to whether you are accepted into a program, it is understandable that taking tests creates anxiety for many students. Even students who know they have learned all the required material find their minds going blank as they stare at the words in the questions. One easy way to overcome that anxiety is to prepare mentally for the test. Mentally preparing for an exam is really not difficult. There are simple techniques that any student can learn to increase their chances of earning a great score on the day of the test.

Do not procrastinate

Study the material for the test when it becomes available, and continue to review the material until the test day. By waiting until the last minute and trying to cram for the test the night before, you actually increase the amount of anxiety you feel. This leads to an increase in negative self-talk. Telling yourself "I can't learn this. I am going to fail" is a pretty sure indication that you are right. At best, your performance on the test will not be as strong if you have procrastinated instead of studying.

Positive self-talk.

Positive self-talk serves both to drown out negative self-talk and to increase your confidence in your abilities. Whenever you begin feeling overwhelmed or anxious about the test, remind yourself that you have studied enough, you know the material and that you will pass the test. Use only positive words. Both negative and positive self-talk are really just your fantasy, so why not choose to be a winner?

Prepare for a Test

Do not compare yourself to anyone else.

Do not compare yourself to other students, or your performance to theirs. Instead, focus on your own strengths and weaknesses and prepare accordingly. Regardless of how others perform, your performance is the only one that matters to your grade. Comparing yourself to others increases your anxiety and your level of negative self-talk before the test.

Visualize.

Make a mental image of yourself taking the test. You know the answers and feel relaxed. Visualize doing well on the test and having no problems with the material. Visualizations can increase your confidence and decrease the anxiety you might otherwise feel before the test. Instead of thinking of this as a test, see it as an opportunity to demonstrate what you have learned!

Avoid negativity.

Worry is contagious and viral - once it gets started it builds on itself. Cut it off before it gets to be a problem. Even if you are relaxed and confident, being around anxious, worried classmates might cause you to start feeling anxious. Before the test, tune out the fears of classmates. Feeling anxious and worried before an exam is normal, and every student experiences those feelings at some point. But you cannot allow these feelings to interfere with your ability to perform well. Practicing mental preparation techniques and remembering that the test is not the only measure of your academic performance will ease your anxiety and ensure that you perform at your best.

How to Take a Test

EVERYONE KNOWS THAT TAKING AN EXAM IS STRESSFUL, BUT IT DOES NOT HAVE TO BE THAT BAD! **There are a few simple things that you can do to increase your score on any type of test. Take a look at these tips and consider how you can incorporate them into your study time.**

Reading the Instructions

This is the most basic point, but one that, surprisingly, many students ignore and it can cost them big time! Since reading the instructions is one of the most common, and 100% preventable mistakes, we have a whole section just on reading instructions.

Pay close attention to the sample questions. Almost all standardized tests offer sample questions, paired with their correct solutions. Go through these to make sure that you understand what they mean and how they arrived at the correct answer. Do not be afraid to ask the test supervisor for help with a sample that confuses you, or instructions that you are unsure of.

Tips for Reading the Question

We could write pages and pages of tips just on reading the test questions. Here are the ones that will help you the most.

- **Think first.** Before you look at the answer, read and think about the question. It is best to try to come up with the correct answer before you look at the options given. This way, when the test-writer tries to trick you with a close answer, you will not fall for it.

- **Make it true or false.** If a question confuses you, then look at each answer option and think of it as a "true" "false" question. Select the one that seems most likely to be "true."

- **Mark the Question.** For some reason, a lot of test-takers are afraid to mark up their test booklet. Unless you are specifically told not to mark in the booklet, you should feel free to use it to your advantage. More on this below.

- **Circle Key Words.** As you are reading the question, underline or circle key words. This helps you to focus on the most critical information needed to solve the problem. For example, if the question said, "Which of these is not a synonym for huge?" You might circle "not," "synonym" and "huge." That clears away the clutter and lets you focus on what is important. More on this below.

- **Always underline these words:** all, none, always, never, most, best, true, false and except.

- **Cross out irrelevant choices.** If you find yourself confused by lengthy questions, cross out anything that you think is irrelevant, obviously wrong, or information that you think is offered to distract you.

- **Do not try to read between the lines.**
Usually, questions are written to be straightforward, with no deep, underlying meaning. The simple answer really is often the correct answer. Do not over-analyze!

How to Take a Test - The Basics

Some tests are designed to assess your ability to quickly grab the necessary information; this type of exam makes speed a priority. Others are more concerned with your depth of knowledge, and how accurate it is. When you receive a test, look it over to determine whether the test is for speed or accuracy. If the test is for speed, like many standardized tests, your strategy is clear; answer as many questions as quickly as possible.

Watch out, though! There are a few tests that are designed to determine how fully and accurately you can answer the questions. Guessing on this type of test is a big mistake, because the teacher expects any student with an average grade to be able to complete the test in the time given. Racing through the test and making guesses that prove to be incorrect will cost you big time!

Every little bit helps

If you are permitted calculators, or other materials, make sure you bring them, even if you do not think you will need them. Use everything at your disposal to increase your score.

Make time your friend

Budget your time from the moment your pencil hits the page until you are finished with the exam, and stick to it! Virtually all standardized tests have a time limit for each section. The amount of time you are permitted for each portion of the test will almost certainly be included in the instructions or printed at the top of the page. If for some reason it is not immediately visible, rather than wasting your time hunting for it you can use the points or percentage of the score as a proxy to make an educated guess regarding the time limit.

Use the allotted time for each section and then move onto the next section whether you have completed the first section or not. Stick with the instructions and you will be able to answer most the questions in each section.

With speed tests you may not be able to complete the entire test. Rest assured that you are not really expected to! The goal of this type of examination is to determine how quickly you can reach into your brain and access a particular piece of information, which is one way of determining how well you know it. If you know a test you are taking is a speed test, you will know the strategies to use for the best results.

Easy does it

One smart way to tackle a test is to locate the easy questions and answer those first. This is a time-tested strategy that never fails, because it saves you a lot of unnecessary fretting. First, read the question and decide if you can answer it in less than a minute. If so, complete the question and go to the next one. If not, skip it for now and continue to the next question. By the time you have completed the first pass through this section of the exam, you will have answered a good number of questions. Not only does it boost your confidence, relieve anxiety and kick your memory up a notch, you will know exactly how many questions remain and can allot the rest of your time accordingly. Think of doing the easy questions first as a warm-up!

If you run out of time before you manage to tackle all the difficult questions, do not let it throw you. All that means is you have used your time in the most efficient way possible by answering as many questions correctly as you could. Missing a few points by not answering a question whose answer you do not know just means you spent that time answering one whose answer you did.

A word to the wise: Skipping questions for which you are drawing a complete blank is one thing, but we are not suggesting you skip every question you come across that you are not 100 % certain of. A good rule of thumb is to try to answer at least eight of every 10 questions the first time through.

Do not watch your watch

At best, taking an important exam is an uncomfortable situation. If you are like most people, you might be tempted to subconsciously distract yourself from the task at hand. One of the most common ways to do so is by becoming obsessed with your watch or the wall clock. Do not watch your watch! Take it off and place it on the top corner of your desk, far enough away that you will not be tempted to look at it every two minutes. Better still, turn the watch face away from you. That way, every time you try to sneak a peek, you will be reminded to refocus your attention to the task at hand. Give yourself permission to check your watch or the wall clock after you complete each section. If you know yourself to be a bit of a slow-poke in other aspects of life, you can check your watch a bit more often. Even so, focus on answering the questions, not on how many minutes have elapsed since you last looked at it.

Divide and conquer

What should you do when you come across a question that is so complicated you may not even be certain what is being asked? As we have suggested, the first time through the section you are best off skipping the question. But at some point, you will need to return to it and get it under control. The best way to handle questions that leave you feeling so anxious you can hardly think is by breaking them into manageable pieces. Solving smaller bits is always easier. For complicated questions, divide them into bite-sized pieces and solve these smaller sets separately. Once you understand what the reduced sections are really saying, it will be much easier to put them together and get a handle on the bigger question.

Reason your way through the toughest questions

If you find that a question is so dense you can't figure out how to break it into smaller pieces, there are a few strategies that might help. First, read the question again and look for hints. Can you re-word the question in one or more different ways? This may give you clues. Look for words that can function as either verbs or nouns, and try to figure out from the sentence structure which it is here. Remember that many nouns in English have several different meanings. While some of those meanings might be related, sometimes they are completely distinct. If reading the sentence one way does not make sense, consider a different definition or meaning for a key word.

The truth is, it is not always necessary to understand a question to arrive at a correct answer! A trick that successful students understand is using Strategy 5, Elimination. Frequently, at least one answer is clearly wrong and can be crossed off the list of possible correct answers. Next, look at the remaining answers and eliminate any that are only partially true. You may still have to flat-out guess from time to time, but using the process of elimination will help you make your way to the correct answer more often than not - even when you don't know what the question means!

Do not leave early
Use all the time allotted to you, even if you can't wait to get out of the testing room. Instead, once you have finished, spend the remaining time reviewing your answers. Go back to those questions that were most difficult for you and review your response. Another good way to use this time is to return to multiple-choice questions in which you filled in a bubble. Do a spot check, reviewing every fifth or sixth question to make sure your answer coincides with the bubble you filled in. This is a great way to catch yourself if you made a mistake, skipped a bubble and therefore put all your answers in the wrong bubbles!

Become a super sleuth and look for careless errors. Look for questions that have double negatives or other odd phrasing; they might be an attempt to throw you off. Careless errors on your part might be the result of skimming a question and missing a key word. Words such as "always", "never", "sometimes", "rarely" and the like can give a strong indication of the answer the question is really seeking. Don't throw away points by being careless!

Just as you budgeted time at the beginning of the test to allow for easy and more difficult questions, be sure to budget sufficient time to review your answers. On essay questions and math questions where you are required to show your work, check your writing to make sure it is legible.

Math questions can be especially tricky. The best way to double check math questions is by figuring the answer using a different method, if possible.

Here is another terrific tip. It is likely that no matter how hard you try, you will have a handful of questions you just are not sure of. Keep them in mind as you read through the rest of the test. If you can't answer a question, looking back over the test to find a different question that addresses the same topic might give you clues.

We know that taking the test has been stressful and you can hardly wait to escape. Just keep in mind that leaving before you double-check as much as possible can be a quick trip to disaster. Taking a few extra minutes can make the difference between getting a bad grade and a great one. Besides, there will be lots of time to relax and celebrate after the test is turned in.

In the Test Room – What you MUST do!

If you are like the rest of the world, there is almost nothing you would rather avoid than taking a test. Unfortunately, that is not an option if you want to pass. Rather than suffer, consider a few attitude adjustments that might turn the experience from a horrible one to...well, an interesting one! Take a look at these tips. Simply changing how you perceive the experience can change the experience itself.

Get in the mood

After weeks of studying, the big day has finally arrived. The worst thing you can do to yourself is arrive at the test site feeling frustrated, worried, and anxious. Keep a check on your emotional state. If your emotions are shaky before a test it can determine how well you do on the test. It is extremely important that you pump yourself up, believe in yourself, and use that confidence to get in the mood!

Don't fight reality

Oftentimes, students resent tests, and with good reason. After all, many people do not test well, and they know the grade they end with does not accurately reflect their true knowledge. It is easy to feel resentful because tests classify students and create categories that just don't seem fair. Face it: Students who are great at rote memorization and not that good at actually analyzing material often score higher than those who might be more creative thinkers and balk at simply memorizing cold, hard facts. It may not be fair, but there it is anyway. Conformity is an asset on tests, and creativity is often a liability. There is no point in wasting time or energy being upset about this reality. Your first step is to accept the reality and get used to it. You will get higher marks when you realize tests do count and that you must give them your best effort. Think about your future and the career that is easier to achieve if you have consistently earned high grades. Avoid negative energy and focus on anything that lifts your enthusiasm and increases your motivation.

Get there early enough to relax

If you are wound up, tense, scared, anxious, or feeling rushed, it will cost you. Get to the exam room early and relax before you go in. This way, when the exam starts, you are comfortable and ready to apply yourself. Of course, you do not want to arrive so early that you are the only one there. That will not help you relax; it will only give you too much time to sit there, worry and get wound up all over again.

If you can, visit the room where you will be taking your exam a few days ahead of time. Having a visual image of the room can be surprisingly calming, because it takes away one of the big 'unknowns'. Not only that, but once you have visited, you know how to get there and will not be worried about getting lost. Furthermore, driving to the test site once lets you know how much time you need to allow for the trip. That means three potential stressors have been eliminated all at once.

Get it down on paper

One advantages of arriving early is that it allows you time to recreate notes. If you spend a lot of time worrying about whether you will be able to remember information like names, dates, places, and mathematical formulas, there is a solution for that. Unless the exam you are taking allows you to use your books and notes, (and very few do) you will have to rely on memory. Arriving early gives to time to tap into your memory and jot down key pieces of information you know will be asked. Just make certain you are allowed to make notes once you are in the testing site; not all locations will permit it. Once you get your test, on a small piece of paper write down everything you are afraid you will forget. It will take a minute or two but by dumping your worries onto the page you have effectively eliminated a certain amount of anxiety and driven off the panic you feel.

Get comfortable in your chair

Here is a clever technique that releases physical stress and helps you get comfortable, even relaxed in your body. You will tense and hold each of your muscles for just a few seconds. The trick is, you must tense them hard for the technique to work. You might want to practice this technique a few times at home; you do not want an unfamiliar technique to add to your stress just before a test, after all! Once you are at the test site, this exercise can always be done in the rest room or another quiet location.

Start with the muscles in your face then work down your body. Tense, squeeze and hold the muscles for a moment or two. Notice the feel of every muscle as you go down your body. Scowl to tense your forehead, pull in your chin to tense your neck. Squeeze your shoulders down to tense your back. Pull in your stomach all the way back to your ribs, make your lower back tight then stretch your fingers. Tense your leg muscles and calves then stretch your feet and your toes. You should be as stiff as a board throughout your entire body.

Now relax your muscles in reverse starting with your toes. Notice how all the muscles feel as you relax them one by one. Once you have released a muscle or set of muscles, allow them to remain relaxed as you proceed up your body. Focus on how you are feeling as all the tension leaves. Start breathing deeply when you get to your chest muscles. By the time you have found your chair, you will be so relaxed it will feel like bliss!

Fight distraction

A lucky few are able to focus deeply when taking an important examination, but most people are easily distracted, probably because they would rather be anyplace else! There are a number of things you can do to protect yourself from distraction.

Stay away from windows. If you select a seat near a window you may end gazing out at the landscape instead of paying attention to the work at hand. Furthermore, any sign of human activity, from a single individual walking by to a couple having an argument or exchanging a kiss will draw your attention away from your important work. What goes on outside should not be allowed to distract you.

Choose a seat away from the aisle so you do not become distracted by people who leave early. People who leave the exam room early are often the ones who fail. Do not compare your time to theirs.

Of course, you love your friends; that's why they are your friends! In the test room, however, they should become complete strangers inside your mind. Forget they are there. The first step is to physically distance yourself from friends or classmates. That way, you will not be tempted to glance at them to see how they are doing, and there will be no chance of eye contact that could either distract you or even lead to an accusation of cheating. Furthermore, if they are feeling stressed because they did not spend the focused time studying that you did, their anxiety is less likely to permeate your hard-earned calm.

Of course, you will want to choose a seat where there is sufficient light. Nothing is worse than trying to take an important examination under flickering lights or dim bulbs.

Ask the instructor or exam proctor to close the door if there is a lot of noise outside. If the instructor or proctor is unable to do so, block out the noise as best you can. Do not let anything disturb you.

Make sure you have enough pencils, pens and whatever else you will need. Many entrance exams do not permit you to bring personal items such as candy bars into the testing room. If this is the case with the exam you are sitting for, be sure to eat a nutritionally balanced breakfast. Eat protein, complex carbohydrates and a little fat to keep you feeling full and to supercharge your energy. Nothing is worse than a sudden drop in blood sugar during an exam.

Do not allow yourself to become distracted by being too cold or hot. Regardless of the weather outside, carry a sweater, scarf or jacket if the air conditioning at the test site is set too high, or the heat set too low. By the same token, dress in layers so that you are prepared for a range of temperatures.

Bring a watch so that you can keep track of time management. The danger here is many students become obsessed with how many minutes have passed since the last question. Instead of wearing the watch, remove it and place it in the far upper corner of the desk with the face turned away. That way, you cannot become distracted by repeatedly glancing at the time, but it is available if you need to know it.

Drinking a gallon of coffee or gulping a few energy drinks might seem like a great idea, but it is, in fact, a very bad one. Caffeine, pep pills or other artificial sources of energy are more likely to leave you feeling rushed and ragged. Your brain might be clicking along, all right, but chances are good it is not clicking along on the right track! Furthermore, drinking lots of coffee or energy drinks will mean frequent trips to the rest room. This will cut into the time you should be spending answering questions and is a distraction in itself, since each time you need to leave the room you lose focus. Pep pills will only make it harder for you to think straight when solving complicated problems on the exam.

At the same time, if anxiety is your problem try to find ways around using tranquilizers during test-taking time. Even medically prescribed anti-anxiety medication can make you less alert and even decrease your motivation. Being motivated is what you need to get you through an exam. If your anxiety is so bad that it threatens to interfere with your ability to take an exam, speak to your doctor and ask for documentation. Many testing sites will allow non-distracting test rooms, extended testing time and other accommodations as long as a doctor's note that explains the situation is made available.

Keep Breathing

It might not make a lot of sense, but when people become anxious, tense, or scared, their breathing becomes shallow and, in some cases, they stop breathing all together! Pay attention to your emotions, and when you are feeling worried, focus on your breathing. Take a moment to remind yourself to breathe deeply and regularly. Drawing in steady, deep breaths energizes the body. When you continue to breathe deeply you will notice you exhale all the tension.

It is a smart idea to rehearse breathing at home. With continued practice of this relaxation technique, you will begin to know the muscles that tense up under pressure. Call these your "signal muscles." These are the ones that will speak to you first, begging you to relax. Take the time to listen to those muscles and do as they ask. With just a little breathing practice, you will get into the habit of checking yourself regularly and when you realize you are tense, relaxation will become second nature.

Avoid Anxiety Before a Test

Manage your time effectively

This is a key to your success! You need blocks of uninterrupted time to study all the pertinent material. Creating and maintaining a schedule will help keep you on track, and will remind family members and friends that you are not available. Under no circumstances should you change your blocks of study time to accommodate someone else, or cancel a study session to do something more fun. Do not interfere with your study time for any reason!

Relax

Use whatever works best for you to relieve stress. Some folks like a good, calming stretch with yoga, others find expressing themselves through journaling to be useful. Some hit the floor for a series of crunches or planks, and still others take a slow stroll around the garden. Integrate a little relaxation time into your schedule, and treat that time, too, as sacred.

Eat healthy

Instead of reaching for the chips and chocolate, fresh fruits and vegetables are not only yummy but offer nutritional benefits that help relieve stress. Some foods accelerate stress instead of reducing it and should be avoided. Foods that add to higher anxiety include artificial sweeteners, candy and other sugary foods, carbonated sodas, chips, chocolate, eggs, fried foods, junk foods, processed foods, red meat, and other foods containing preservatives or heavy spices. Instead, eat a bowl of berries and some yogurt!

Get plenty of ZZZZZZZs

Do not cram or try to do an all-nighter. If you created a study schedule at the beginning, and if you have stuck with that schedule, have confidence! Staying up too late trying to cram in last-minute bits of information is going to leave you exhausted the next day. Besides, whatever new information you cram in will only displace all the important ideas you've spent weeks learning. Remember: You need to be alert and fully functional the day of the exam

Have confidence in yourself!

Everyone experiences some anxiety when taking a test, but exhibiting a positive attitude banishes anxiety and fills you with the knowledge you really do know what you need to know. This is your opportunity to show how well prepared you are. Go for it!

Be sure to take everything you need

Depending on the exam, you may be allowed to have a pen or pencil, calculator, dictionary or scratch paper with you. Have these gathered together along with your entrance paperwork and identification so that you are sure you have everything that is needed.

Do not chitchat with friends

Let your friends know ahead of time that it is not anything personal, but you are going to ignore them in the test room! You need to find a seat away from doors and windows, one that has good lighting, and get comfortable. If other students are worried their anxiety could be detrimental to you; of course, you do not have to tell your friends that. If you are afraid they will be offended, tell them you are protecting them from your anxiety!

Common Test-Taking Mistakes

Taking a test is not much fun at best. When you take a test and make a stupid mistake that negatively affects your grade, it is natural to be very upset, especially when it is something that could have been easily avoided. So what are some of the common mistakes that are made on tests?

Do not fail to put your name on the test

How could you possibly forget to put your name on a test? You would be amazed at how often that happens. Very often, tests without names are thrown out immediately, resulting in a failing grade.

Marking the wrong multiple-choice answer

It is important to work at a steady pace, but that does not mean bolting through the questions. Be sure the answer you are marking is the one you mean to. If the bubble you need to fill in or the answer you need to circle is 'C', do not allow yourself to get distracted and select 'B' instead.

Answering a question twice

Some multiple-choice test questions have two very similar answers. If you are in too much of a hurry, you might select them both. Remember that only one answer is correct, so if you choose more than one, you have automatically failed that question.

Mishandling a difficult question

We recommend skipping difficult questions and returning to them later, but beware! First, be certain that you do return to the question. Circling the entire passage or placing a large question mark beside it will help you spot it when you are reviewing your test. Secondly, if you are not careful to skip the question, you can mess yourself up badly. Imagine that a question is too difficult and you decide to save it for later. You read the next question, which you know the answer, and you fill in that answer. You continue to the end of the test then return to the difficult question only to discover you didn't actually skip it! Instead, you inserted the answer to the following question in the spot reserved for the harder one, thus throwing off the remainder of your test!

Incorrectly Transferring an answer from scratch paper

This can happen easily if you are trying to hurry! Double check any answer you have figured out on scratch paper, and make sure what you have written on the test itself is an exact match!

Don't ignore the clock, and don't marry it, either

In a timed examination many students lose track of the time and end up without sufficient time to complete the test. Remember to pace yourself! At the same time, though, do not allow yourself to become obsessed with how much time has elapsed, either.

Thinking too much

Oftentimes, your first thought is your best thought. If you worry yourself into insecurity, your self-doubts can trick you into choosing an incorrect answer when your first impulse was the right one!

Be prepared

Running out of ink and not having an extra pen or pencil is not an excuse for failing an exam! Have everything you need, and have extras. Bring tissue, an extra eraser, several sharpened pencils, batteries for electronic devices, and anything else you might need.

Not following directions

Directions are carefully worded. If you skim directions, it is very easy to miss key words or misinterpret what is being said. Nothing is worse than failing an examination simply because you could not be bothered with reading the instructions!

Conclusion

CONGRATULATIONS! You have made it this far because you have applied yourself diligently to practicing for the exam and no doubt improved your potential score considerably! Passing your up-coming exam is a huge step in a journey that might be challenging at times but will be many times more rewarding and fulfilling. That is why being prepared is so important.

Good Luck!

Register for Free Updates and More Practice Test Questions

Register your purchase at https://www.test-preparation.ca/register/ for fast and convenient access to updates, errata, free test tips and more practice test questions.

Online Resources

How to Prepare for a Test - The Ultimate Guide

https://www.test-preparation.ca/prepare-test/

Learning Styles - The Complete Guide

https://www.test-preparation.ca/learning-style/

Test Anxiety Secrets!

https://www.test-preparation.ca/test-anxiety/

Time Management on a Test

https://www.test-preparation.ca/time-management/

Flash Cards - The Complete Guide

https://www.test-preparation.ca/flash-cards/

Test Preparation Video Series

https://www.test-preparation.ca/test-video/

How to Memorize - The Complete Guide

https://www.test-preparation.ca/memorize/

www.ingramcontent.com/pod-product-compliance
Lightning Source LLC
Chambersburg PA
CBHW072014070526
44583CB00015B/1481